P4
£5.95

Geography matters!

Geography matters!
A reader

EDITED BY
DOREEN MASSEY AND JOHN ALLEN

WITH JAMES ANDERSON, SUSAN CUNNINGHAM,
CHRISTOPHER HAMNETT AND PHILIP SARRE

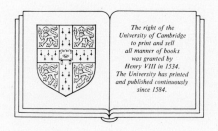

The right of the
University of Cambridge
to print and sell
all manner of books
was granted by
Henry VIII in 1534.
The University has printed
and published continuously
since 1584.

CAMBRIDGE UNIVERSITY PRESS

CAMBRIDGE

LONDON   NEW YORK   NEW ROCHELLE

MELBOURNE   SYDNEY

IN ASSOCIATION WITH
THE OPEN UNIVERSITY

Published by the Press Syndicate of the University of Cambridge
The Pitt Building, Trumpington Street, Cambridge CB2 1RP
32 East 57th Street, New York, NY 10022, USA
296 Beaconsfield Parade, Middle Park, Melbourne 3206, Australia

First published 1984

Printed in Great Britain by the University Press, Cambridge

Library of Congress catalogue card number: 84-16969

British Library Cataloguing in Publication Data

Geography matters!
1. Anthropo-geography
I. Massey, Doreen B.   II. Allen, John
304.2   GF41

ISBN 0 521 26887 7 hard covers
ISBN 0 521 31708 8 paperback

Contents

List of Acknowledgements

The editors would like to thank members of the course team of 'Changing Britain, Changing World: Geographical Perspectives' for their help in selecting material for this reader. We would like to thank the authors of the chapters, especially those which were specially commissioned, for their receptiveness to our comments and cuts and for their adherence to a tight production schedule. We would also like to thank Maureen Adams, Eve Hussey, Enid Sheward and Giles Clark for assistance in turning manuscripts and defaced xeroxes into a legible typescript.

The Press is grateful to the authors and publishers listed below for permission to reproduce copyright material from the following works:

Chamberlain, M. Extracts from *Fenwomen* (London, Routledge and Kegan Paul 1983).

Harrison, Paul. Extracts from pp. 64–68 of *Inside the Inner City* (London, Pelican Books 1983). Copyright © Paul Harrison, 1983, and reprinted by permission of Penguin Books Ltd.

Kolinsky, Martin. 'The Nation-State in Western Europe: Erosion from Above and Below' in L. Tivey (ed.), *The Nation State*, pp. 82–103 (Oxford, Martin Robertson 1981).

Murgatroyd, Linda and John Urry. 'The Restructuring of a Local Economy: the Case of Lancaster' in J. Anderson, S. Duncan and R. Hudson (eds.), *Redundant Spaces in Cities and Regions*, pp. 67–96 (London, the Academic Press 1983). Copyright © 1983, By Academic Press Inc.

Piciotto, Sol. 'Jurisdictional Conflicts, International Law and the International State System' in *International Journal of the Sociology of Law*, Vol. 11, 1983, pp. 11–40 (Copyright 1983 by Academic Press Inc (London) Ltd).

Richards, Alan and Philip L. Martin. 'The Laissez-Faire Approach to International Labor Migration: The Case of the Arab Middle East' in *Economic Development and Cultural Change* Vol. 31 no 3, April 1983, pp. 455–471 (© University of Chicago Press, 1983. All rights reserved).

Sack, Robert David. *Conceptions of Space in Social Thought: A Geographic Perspective*, pp. 167–193. (Macmillan, London and Basingstoke, and the University of Minnesota Press, 1980).

Sandbach, Francis. 'Environmental Futures' in *Environment, Ideology and Policy*, pp. 200–223 (Oxford, Basil Blackwell 1980).

PART 1

Introduction
Geography matters

DOREEN MASSEY

Insofar as any sensible distinction can be made between the various social science disciplines, 'human geography' has traditionally been distinguished by its concern with three relationships. First, there is the relationship between the social and the spatial: between society and social processes on the one hand and the fact and form of the spatial organization of both of those things on the other. Second, there is the relationship between the social and the natural, between society and 'the environment'. Third, there is a concern, which geography shares in particular with history, with the relationship between different elements – economy, social structure, politics, and so forth. While the 'substance' disciplines of the social sciences (economics, sociology, politics) tend to focus on particular parts of society, however difficult these are to distinguish and define, human geography's concern with 'place', with why different localities come to be as they are, has often led it to the study of how those different elements come together in particular spaces to form the complex mosaic which is the geography of society.

The way in which each of these relationships has been conceptualized has varied widely, and often quite dramatically, even in the recent history of the discipline. All have had their extreme versions. The most absolute of environmental determinists saw human character and social organization as a fairly direct and unmediated product of the physical (natural) environment. Some of the 'models of spatial interaction' in the era of the quantitative nineteen-sixties posited a realm of 'the spatial' virtually as substantive as the economic is for economists. There have been studies in which the 'synthesis of elements' (after which regional geography supposedly strove) amounted to little more than chapters which began with geology and gradually moved 'upwards' to politics and culture, with little attempt at interlinkage let alone theorization. But the fact that the answers have so often been wrong does not mean that the questions which were being addressed were not significant. Indeed, what we want to argue in this book is that these questions, concern for these relationships, are of great importance, not just for 'human geography' but for the social sciences as a whole, and for what those social sciences are all about – understanding, and changing, society.

It is our intention here to argue for particular interpretations of each of

1

these relationships. Much of the debate within social science in the sixties and seventies was in fact concerned, if only implicitly, with these issues. Certainly, the debates had implications for social science's attitude to each of the relationships highlighted here. What we want to do now is to make those relationships explicit, and then to argue that on each front the debate needs to be pushed forward another step. The interpretations which we argue for here are intended to do just that. They enable both 'the spatial' and 'the natural' to regain a significance, which they had previously lost, within the social sciences as a whole. And the concern for place, and hence for specificity and uniqueness, has its parallels, and hence also its implications, in some of the central methodological debates in the social sciences today.

The social and the spatial

One of the classic images of human geography, perhaps today more amongst those outside it than those within, is its concern for the region, the area, the locality, call it what you will. Most of the editors of and contributors to this book were educated in a school-geography which divided itself up into courses on particular regions. This was frequently paralleled by a 'systematic geography' in which the spatial organization of different elements (population, industry, disease) was studied across a set of regions. But it was regional geography which was the central focus. And in that regional geography the concern with space was bound up with a recognition of uniqueness and specificity. Each place was different, and the aim was to put together the elements in such a way that each configuration could be understood. The problem with that project was not its aim. Indeed we are going to argue here that that aim should once again be far more prominent on our agenda. The problem was its execution. Too often it degenerated into an essentially descriptive and untheorized collection of facts. It was this period that gave geography its name as the discipline where you learned lists of products.

The intellectual weakness of this tradition by the middle of the twentieth century laid human geography open to the wave of super-positivism and the mania for quantification which swept all the social sciences in the nineteen-sixties. Geography was to become 'scientific', in the strictly positivist sense. The apogee of this attempt was to be found in the school of 'spatial analysis'. Here mathematical models were built of 'spatial interactions', theories and general laws were constructed out of empirical generalizations from vast data-sets, and everything that was conceivably quantifiable was quantified. In the process much was lost. Most obviously, things which would not submit so easily to quantification disappeared from view. In parallel fashion 'space' itself was reduced to a concern with distance; the interest in particularity and uniqueness was replaced by a search for spatial regularities. But the other thing that was lost was geography's own distinctiveness. It had lost its old distinctive focus – the synthesis of elements within the individuality of a particular region. And its methods had converged with those of the other sciences. Geography might have, in the terms of the sixties, made itself into a science. The question was: a science of what?

The answer given was: a science of the spatial. The models of spatial interaction posited, either implicitly or explicitly, the notion of 'purely spatial processes'. 'Spatial effects' (the geographical distribution of one thing) were deemed to be explicable by 'spatial causes' (the geographical distribution of another). During this period human geography carved out for itself a new distinctiveness by defining a new object of study – the realm of the spatial. There were spatial laws and spatial processes, spatial causes and spatial relationships. Nor was this an argument whose implications were confined to the intellectual or the academic. It was an important element in the debate over the causes of inner-city decline, and over the impact of regional policy.

In terms of the relation between the social and the spatial, this was the period of perhaps the greatest conceptual separation. Geography at this time, or at least the dominant school of geographical thought, had separated off for itself a realm of the spatial, self-containedly including, in this conception, both cause and effect. It needed little input from the other social science disciplines. For their part the other disciplines forgot about space altogether.

It could not last. In the nineteen-seventies, again along with the other social sciences, a radical critique was launched in human geography of the dominant school of the sixties. In geography it took a particular form. Above all it was argued – and it now seems so obvious – that there is not and cannot be a separate realm of 'the spatial'. There are no such things as spatial processes without social content, no such things as purely spatial causes, spatial laws, interactions or relationships. What was really being referred to, it was argued, was the spatial form of social causes, laws, interactions and relationships. 'The spatial', it was pronounced, and quite correctly, 'does not exist as a separate realm. Space is a social construct.'

Once again, this had wider ramifications than the simply academic, and indeed it was most frequently in debates about policy issues that the methodological questions came most clearly to a head. The causes of spatial patterns, such as inner-city decline and the problems of peripheral regions, could not be sought simply in other spatial patterns, it was argued; the causes had to be found in wider changes going on in British economy and society as a whole.

One of the immediate implications of this argument was that, in order to explain their chosen object of study, geographers now had to go outside what had previously been conceived of as the boundaries of their discipline. In order to understand the geography of industry it was necessary to learn economics and industrial sociology. To understand spatial differentiation in housing it was necessary to appreciate the mechanisms (economic, sociological, political) underlying the operation of the housing market. It was recognized, in other words, that in order to understand 'geography' it was necessary to understand society.

The position of this book is to argue that the critique was both correct and important. Indeed, it seems difficult in retrospect to do otherwise. But we also want to argue that this is not the end of the debate.

The position reached in the argument so far is inadequate in a number of related respects. Essentially, only one half of the argument had been followed through. It had been agreed that the spatial is a social construct. But the corollary, that social processes necessarily take place over space, had not been taken on board. While geographers struggled to learn other disciplines and apply their knowledge to the understanding of spatial distributions, the other disciplines continued to function, by and large, as though the world operated, and society existed, on the head of a pin, in a spaceless, geographically undifferentiated world. In terms of the academic disciplinary division of labour, this left geographers simply mapping the outcomes of processes studied in other disciplines; the cartographer of the social sciences. And much of the 'radical geography' of the period was indeed of the 'mapping poverty' variety. But this unsatisfactory division of labour reflected a far more important problem at the level of conceptualisation. For 'space' was seen as only an outcome; geographical distributions as only the *results* of social processes.

But there is more to it than that. Spatial distributions and geographical differentiation may be the result of social processes, but they also affect how those processes work. 'The spatial' is not just an outcome; it is also part of the explanation. It is not just important for geographers to recognize the social causes of the spatial configurations that they study; it is also important for those in other social sciences to take on board the fact that the processes they study are constructed, reproduced and changed in a way which necessarily involves distance, movement and spatial differentiation.

Two questions arise immediately from this formulation. First, what does it mean to say that space has effects? One thing it does *not* mean is that 'space itself', or particular spatial forms, themselves have effects. That would simply be to reproduce the mistakes of the sixties and to posit a notion of the purely spatial. To take an example: it is frequently argued these days that what is emerging as industry's new spatial pattern in the UK will pose insuperable problems for trade union organizers. It is a pattern of relatively isolated factories, often in small towns, often with one plant dominating a whole labour market. Contrasts are drawn with the industrially mixed large labour markets and socially rich contexts of the cities, where trades councils and militancy have frequently flourished. With foreboding (if you are in favour of active trade union organization), people point to the paternalistic relations, lack of militancy, and the frequent common front adopted by workers and owner-managers in the small and isolated labour markets which have so often in the past typified, for instance, the old textile industry. Spatial form seems to 'explain' the difference between the two: city versus isolated labour market. But it does not. It is only necessary to think of the colliery villages of the old coalfields, equally small labour markets, equally dominated by a single employer, for the argument to fall. For the colliery villages on numbers of occasions have been centres of radicalism and militancy. So indeed have the cotton towns, in their day, been hotbeds of radicalism. It is not spatial form in itself (nor distance, nor movement) that has effects, but the spatial form of particular and specified social processes and social relationships. The social character of both capital and labour has been vastly different in the textile towns and the colliery villages and in both it has changed over time. In both, spatial form has indeed been important in terms of trade union organization and militancy (or lack of it) but it has been a spatial form of different social relationships (with different social content) and as such its influence has been different.

The second question is what, anyway, do we mean by 'space'? As we have already seen, the answer has varied over time. The 'old regional geography' may have had its disadvantages but at least it did retain within its meaning of 'the spatial' a notion of 'place', attention to the 'natural' world, and an appreciation of richness and specificity. One of the worst results of the schools of quantification and spatial analysis was their reduction of all this to the simple (but quantifiable) notion of distance. Space became reduced to a dimension. The arguments of the seventies, by reducing the importance of the spatial, downplayed also any explicit debate about its content. In our view, the full meaning of the term 'spatial' includes a whole range of aspects of the social world. It includes distance, and differences in the measurement, connotations and appreciation of distance. It includes movement. It includes geographical differentiation, the notion of place and specificity, and of differences between places. And it includes the symbolism and meaning which in different societies, and in different parts of given societies, attach to all of these things.

All these aspects of 'the spatial' are important in the construction, functioning, reproduction and change of societies as a whole and of elements of society. *Distance* and separation are regularly used by companies to establish degrees of monopoly control, whether it be over markets (the corner shop being the classic – though possibly least important – example) or over workers (the great advantage for capital of those colliery and textile villages was that, short of migration, workers had no alternative place to sell their labour). *Movement*, and more generally locational flexibility, has become in recent years a major weapon used by capital against labour. Threats to close and move elsewhere have become an almost automatic response of big firms faced with resistance from labour. And in times of recession and shortage of jobs it is a powerful threat. More generally, the search after cheaper labour in recent decades has precisely involved spatial movement, whether it be internationally (with the building of a new international division of labour) or intranationally, with the decentralization of production to 'the regions' of Britain in the sixties and seventies. In both cases spatial restructuring was integral to the maintenance of profitability. A sense of *place*, a commitment to location and to established community, can be a strong element of people's resistance to planners' plans. Notions of place and territory are fundamental elements of state politics. And the *symbolism* of space and place, which varies both between societies and within them, from the landscapes of Aboriginal Dreamtime to 'prime sites' and 'prestige locations' for bank headquarters, to the Lincoln Memorial and the Cenotaph, in all these forms is integral to the mode, and effectiveness, of social organization.

It is not just that the spatial is socially constructed; the social is spatially constructed too.[1]

The social and the natural

The logic of our argument about the relationship between the social and the natural is similar to that about the social and the spatial. Indeed the two are closely related: the symbolism of place is often related to natural features, questions of space are intimately bound up with notions of territory and thence of land, part of the uniqueness of places is a result of physical characteristics, of landforms perhaps, or climate.

There is also once again a history to the way in which geography and geographers have conceptualized the relationship. And once again, too, it is bound up with debates within the social sciences more generally.

Probably the most notorious school of thought upon this subject has been that of environmental determinism, with its view that an important explanation of the way in which society is organized, and human beings behave, is the natural environment. It is a school which in its extreme and developed form faded from the forefront of geographical thought many decades ago. It is important to mention it now because it left a legacy. It was a legacy which took many forms: that natural wealth, richness in natural resources, was responsible for economic development, the physical dereliction of inner cities for the destitution of their inhabitants, that 'natural' causes – drought, flood, crop failure – were responsible for famine, hunger and poverty in large parts of the world.

The school which challenged environmental determinism toned down but by no means eradicated that legacy. This was the school of 'possibilism'. It was an inelegant but accurate title, for what the possibilists argued was that nature had to be seen not as determining social action, but as providing a set of options and constraints. You cannot mine coal where none exists, but the existence of coal does not spontaneously produce a mine. 'Society decides.' Yet this view, in spite of its differences from environmental determinism, also shared something with it. For both schools, the way in which nature was conceptualized was unproblematical; its physical reality was simply evident.

But recent decades, again, have seen major critiques both of this way of conceptualizing nature and of the remnants of the determinist view. Indeed in many ways the new argument sought to turn determinism on its head. It was designed to combat the notion of unmediated natural cause. Quite correctly it pointed to the social causes of famine, to the social articulation of what the media announced on the news as 'natural disasters', to the fact that the availability or otherwise of resources was a social question, that while on the one hand the cry was going up that resources were running out, on the other hand coal mines, for instance, were being closed up with good coal still within them. It was essentially an optimistic critique, and it was important. Phenomena which are the product of society are changeable. The natural – just like the spatial – is socially constructed.

But – as in the case of the spatial – the critique went too far and threw out too much. Instead of a real reconceptualization it made the social all-powerful and eradicated nature. But the social is not all there is: social relations are constructed in and as part of a natural world. Nor is it simply a question of options and constraints, as the possibilists would have it. For that is to posit again the notion of two separate spheres. We can only think of the social conquering the natural, or of the natural presenting constraints to the social, if the two spheres are initially assumed to be separate.

Once again conceptualization is central. Ideas of nature, just like those of space, have changed dramatically through history. There have been contrasts between societies and conflicts within them over what should be the dominant view. And those conflicts have been, and are, more than intellectual alter- cations; they reflect struggles over the organization of society and over what should be its priorities. The emergence of capitalism brought with it enormous changes in the dominant view of nature; from animate Mother Earth, to source of resources and profit, to the endlessly cataloguable and improvable. The geographical expansion of capitalism was often viewed in terms of putting resources to better use, gaining greater control over nature (in fact gaining greater control over other societies, and other views of nature). Conflicts between mining companies and Australian Aborigines today embody the same kinds of confrontation – the earth as source of profit from uranium or as sacred sites from time immemorial. Planning enquiries over proposed new coalfields in the midlands of England bring into confrontation notions of land as landed property, nature as resources and 'the natural' as a weekend escape. In other parts of the world, the peasants' positive use of the variety and multiple richness of nature comes up against the logic of commercial agriculture's desire to eradicate that unpredictability and richness, to control it through the application of 'science', to grow crops in endless landscapes of monoculture, to put chickens into factories, and to produce the square tomato.

To face these issues, it is necessary to go beyond the critique of the seventies.

As we said, then, the logic of this theme is similar to that of the relation between the social and the spatial. Both start from the rejection of simply autonomous spheres. In the case of the relation between society and nature, the argument starts, certainly, by rejecting the notion of an unmediated effect of nature on society: that famine is simply the result of natural conditions, for instance, or development the result of natural resources. Natural materials are not even necessarily natural resources; certain social conditions are necessary for them to become so. Within the social sciences, if not within the world of day-to-day politics and the public media, this is clearly now a widely accepted position. But to accept that position does not mean that the world is in some sense 'totally social'. This has its implications both for the conceptualization of social processes within the social sciences and for society itself. On the one hand if we can only conceptualize 'the natural' through the prism of the social so also we need to be aware when analysing social processes that they necessarily take place within a 'natural' world. On the other hand, we recognize that social processes have effects upon the natural environment – it is an impact which, in the industrialized world, is often phrased in terms such as conquering or controlling. All such terms give the impression that society is in charge, and some of the consequences of that view are becoming increasingly apparent. From acid rain to potential climatic disaster through devastation of the world's major forests, 'nature' is hitting back. It is clear that the conceptualization of the social and the natural as two separate spheres, and in particular the variety of views (from all parts of the political spectrum) that in that duality the social (frequently manifesting itself in the noun 'man') controls the natural, is inadequate. A lot depends on our recognizing that neither 'the social' nor 'the natural' can be conceptualized in isolation from the other.

Uniqueness and interdependence

There is another way, too, in which the argument we are presenting here attempts to push forward the state of debate in the social sciences generally. Any consideration of geography in the fullest sense of the word must face up to the theoretical problem of the analysis of the unique. In one sense the very thing that we study is variation: each place is unique. This, too, is something which has been lost sight of in the social science debate of recent years, in the search after general laws, the intellectual dominance of certain forms of 'top-down' structuralism, the (quite correct) desire to relate the individual occurrence to the general cause.

It was fundamentally important to argue in the nineteen-seventies that inner-city decline stemmed from general processes of deindustrialization in the British economy, indeed from the reorientation of the position of that economy within the shifting international division of labour. It was important to counter the then prevailing orthodoxy that the explanation for the problems of the inner cities could be found within the inner cities themselves. It was important, in other words, to show how the specific outcomes (the collapse of Merseyside, the London docks, central Glasgow) were all the product of more general causes.

And yet, too, in making that point so strongly, something was sacrificed – the importance of specificity, the ability to explain, understand, and recognize the significance of, the unique outcome.

The fundamental methodological question is how to keep a grip on the generality of events, the wider processes lying behind them, without losing sight of the individuality of the form of their occurrence. Pointing to general processes does not adequately explain what is happening at particular moments or in particular places. Yet any explanation must include such general processes. The question is how. Too often a solution has been sought through an uneasy, and untenable, juxtaposition of two kinds of explanation. On the one hand, the 'general', whether it be in the form of immanent tendencies or empirically identified wider processes, is treated in deterministic fashion. On the other hand, since the infinite variety of reality does not in fact conform to this logic, additional factors are added on, in ad hoc and descriptive fashion, to explain (explain away) the deviation.

But variety should not be seen as a deviation from the expected; nor should uniqueness be seen as a problem. 'General processes' *never* work themselves out in pure form. There are always specific circumstances, a particular history, a particular place or location. What is at issue – and to put it in geographical terms – is the articulation of the general with the local (the particular) to produce qualitatively different outcomes in different localities. To take an example: the decentralization of 'women's jobs' has taken place in recent decades in the United Kingdom to a whole range of regions – to East Anglia, South Wales, Cornwall. But the impact of that decentralization (the result, the outcome) has been different in each place. Each region was distinct (unique) before the process took place, and in each place the local conditions/ characteristics operated on the general processes to produce a specific outcome. In each case uniqueness was reproduced, and in each case it was also changed. If this is in some sense 'structural analysis', it is in no sense simply 'top-down'.

This issue is important. Most obviously it is important to be able rigorously to explain particularity. Only then is it possible to understand a society as it is, in its specific form and with its internal variations. But it is also that this specificity is in turn important *in* explanation. In 'geographical questions' this is so in a number of ways: as we have just argued, regional specificity has an impact on the operation of general, national or international level processes, for instance. And the whole mosaic of regional specificities, the fact of geographical variety itself – in the labour movement, in unemployment rates, in political traditions – can have an enormous impact on the way that society 'as a whole', at national level, is reproduced and changed. These examples are taken from human geography and relate to one of its central concerns: the fact of uneven development and of interdependent systems of dominance and subordination between regions on the one hand, and the specificity of place on the other. It is in this form that the problem of the general and the unique most clearly presents itself to geography. It is a problem which has been around for some time. As we saw, the positivist spatial scientists threw out the unique as unamenable to anything but description. The radical critique recognized it but saw the most important task to be linking the specific to the general. That task is still important. But it is also necessary to reassert the existence, the explicability, and the significance, of the particular. What we do here is take up again the challenge of the old regional geography, reject the answers it gave while recognizing the importance of the problem it set, and present our own, very different solution.

The structure of the book

Our overall theme, then, is that geography, in the fullest sense of the word, matters. This argument is presented in different ways in each of the sections which follow.

The first two pieces are basic building blocks. They address head on the question of conceptualization, considering in turn the two key terms of our discussion: nature and space. We have already argued that the conceptualization of each has varied both between societies and within them. Mick Gold, in his consideration of 'the history of nature', and Robert Sack in his discussion of 'societal conceptions of space' each trace out elements and aspects of this variation. Mick Gold concentrates on variations through time in European conceptualizations of nature; while Robert Sack contrasts views of space in what he terms respectively 'primitive' and 'civilized' societies. From both pieces two things are clear: that the issue of conceptualization is bound up with social form and social order, and that in both maintaining and challenging social orders different forms of conceptualization of both space and nature are frequently at stake.

Mick Gold's article in effect encapsulates the whole of our argument about the relation between the social and the natural. As he makes clear, the issue of what is the socially dominant conceptualization of that relation has more than an intellectual significance. It is also utterly practical. What is needed is a theory of social change which fully incorporates the fact of its existence as an integral part of a physical world, and social practices which work within that knowledge.

Sack, too, presents powerful arguments relating conceptions of space to the internal organization of society. In 'primitive' societies, he argues, the organic relationship between the individual and society is reflected in the relationship between society and milieu. The emergence of 'civilized' societies involves the separation of these terms, and their attempted recombination. In that context, the territorial definition of society becomes central to the maintenance of control. Sack makes a point perhaps too often overlooked: that 'Whatever else a state may be or do, it is territorial.' A challenge to territory is a challenge to the state. In other ways, too, from territorial control over the workplace, to its use of the symbolism of national shrines and holy places, civilized society requires and uses particular, and varied, conceptions of space and place.

It is not just that geography, in the sense of space and nature, matters, but that the way in which we conceptualize those terms in the first place is crucial too.

These two articles follow immediately after this Introduction. The rest of the book is divided into three further sections. In the first of these: 'Analysis: aspects of the geography of society', we examine a set of different individual elements of the functioning of society, particular social processes, or bundles of social processes. The ones we have chosen concern the urban economy, cultural forms and international law. They are deliberately varied, because our aim is to show how spatial structure and changes in geographical organization are important to the functioning of a wide range of social processes. The next section: 'Synthesis: interdependence and the uniqueness of place', is where we address the question of specificity, the problem of relating the general and the unique. We pose it in the context of understanding uneven development – a context where interdependence and uniqueness are two sides of the same coin – and present an alternative answer to the challenge posed by the old regional geography. In the final section: 'Geography and society', we address the issue at the broadest level – why does 'geography' matter to the functioning of society as a whole? This can be tackled at a variety of levels and from a number of angles. Some of them will already have become obvious from discussions earlier in the book. In this last section, however, we have decided to address two major questions for the future – the challenges, from above and below, to the fundamental political territorial unit of the modern world, the nation state; and the question of what threats to society have arisen from the presently dominant forms of relation between the social and the natural: what is the relation between the prospects for ecodoom and the way in which society is organized? Both of these major questions have at their centre the fact that society is part of a world in which both 'the spatial' and 'the natural' are fundamental components.

Notes

1 These arguments are developed further in Massey, D. B. (1984) *Spatial Divisions of Labour: social structures and the geography of production*, Macmillan; and in Massey, D. B. (1984) 'New directions in space' in Urry, J. and Gregory, D. (eds.), *Social Relations and Spatial Structures*, Macmillan.

1

A history of nature

MICK GOLD

'Nature' is a complicated word: it has different meanings and these meanings affect each other.[1] To discover how nature has been represented in Western culture, it may help to distinguish three very basic meanings: (1) the essential quality or character of something (the corrosive nature of salt water); (2) the underlying force which directs the world (nature is taking her course); (3) the material world itself, often the world that is separate from people and human society (to re-discover the joys of nature at the weekend).

If nature is usually taken as referring to the world 'out there' – from the smallest grasshopper, through the Grand Canyon, to the most distant galaxy – it is also believed that there is a force at work, that nature is working according to certain principles and that if we study nature we can deduce a moral lesson.

And that is why nature has a history. Nature cannot simply be regarded as what is out there – a physical universe which preceded the world of human values, and which will presumably outlive the human race – because what is out there keeps changing its meaning. Every attempt at describing nature, every value attributed to Nature – harmonious, ruthless, purposeful, random – brings nature inside human society and its values.

This essay is about how views of nature have changed through the centuries, and what these changes tell us about human history. It concentrates on European accounts of nature from the late Middle Ages until the present day, and the ways in which the personality of nature has been affected by social forces. Every time we use the word 'natural' ("It's only natural, isn't it?") we are almost unconsciously referring to a world order which can be invoked to justify or to make sense of what happens in this world: from a strong sexual attraction ("Doing what comes naturally") to the devastation of famine ("A cruel natural catastrophe"). There are several different versions or traditions of nature present in Western culture. This essay is a brief introduction to the social forces that have expressed, and even altered, the personality of nature.

The animate earth

It's worth noticing that Western culture refers to a singular force called 'Nature' which is frequently personified as a woman: Dame Nature in mediaeval mystery plays; Mother Nature in everyday speech. Raymond Williams has pointed out that in other cultures, nature is represented as a complex network of forces: a spirit of the rain, a spirit in the wind, and so on. The Western model of nature – singular and often personified – is connected with the fact that we live in a culture based on monotheism.[2] And in mediaeval culture, Nature was described as God's deputy – responsible for carrying out His wishes on Earth. One may speculate that this relationship embodied a compromise between Christianity and an older, animistic way of interpreting the world: a way of seeing the Earth as a living body, of seeing live forces in rocks and winds as well as in trees and animals. Although this animistic vision was superseded by the Christian world view and then by scientific values, it is extremely tenacious and clings on to us through figures of everyday speech: even today it is common to describe the way in which farmers allow fields to lie fallow for a year as "giving the soil a chance to rest". Such feelings are a reminder of the days when the whole world was felt to be alive.

In Renaissance maps, it was conventional to represent the wind as gusts of breath which issued from the mouths of cherubs and angels. This was but one element of the way in which the Earth was perceived as a living body: the circulation of water through rivers and seas was comparable to the circulation of blood; the circulation of air through wind was the breath of the planet; volcanoes and geysers were seen as corresponding to the Earth's digestive system – eruptions were like belches and farts issuing out of a central stomach.

Nature was not a process 'out there' to be analysed or exploited. Within the Scholastic world view (the mediaeval synthesis of Aristotelian and Christian doctrines), nature was a complex chain connecting God to the humblest pebble; and man[3] was emphatically inside the system: above the other creatures by virtue of his reason, but below the angels.[4] This belief system was not just about nature as we understand it today; it also embraced the structure of society.

Descriptions of the state talked of the "body politic" in which the princes were the rulers – the brain – and the Church was the soul. The distribution of wealth was the job of the financial system, just as the stomach and intestines were responsible for the distribution of food. And the feet upon which society rested were the peasants and artisans. It was a very hierarchical model, but one which stressed the interdependence of all the elements.[5]

Correspondences not only existed between our bodily functions and the social order; they also existed between the human body and the universe: this is the view that underlies astrology. Each part of the body was thought to be governed by a zodiacal sign, so that the whole body was a miniature replica (or microcosm) of the celestial spheres. Furthermore, the personality of each man and woman was said to be described by the positions of the stars at the moment of birth, so that we all carry within ourselves a model of the universe.

Through this system of correspondences, man: society: the universe were unified into a coherent system. A system that was replaced by the rise of a scientific outlook which broke the individual, society, and nature down into their component parts to explain how the bits fitted together. In 1500 the dominant image of nature and the world was the living organism. By 1700 the dominant image was the machine.

Natural resources

The change from one outlook to another was not a clean break; for some two hundred years these two systems overlapped. There was confusion about what was alive and what wasn't. When Harvey gave the first clear account of the circulation of the blood (1628), many geologists were excited: they thought this was the way in which water circulated the planet. If blood can flow uphill, then why not water? The Jesuit scholar, Athanasius Kircher, was one of the first to study volcanoes and geysers at close quarters. In the pictures that he drew of these phenomena, we can see the beginnings of a scientific study, but also the belief that water flowed around the world through a circulatory system; and his pictures of mountains conveyed his view that they could be the skeletons of gigantic animals.[6]

This is the picturesque side of a change in world view, but there was also a moral and political dimension. If the world were thought of as alive, this placed some restraints on how the world was treated. We are familiar with this idea from other cultures; we know that American Indians regarded the Earth as a great mother and found the white man's way of exploiting the Earth abhorrent:

You ask me to plow the ground! Shall I take a knife and tear my mother's breast? You ask me to dig for stones! Shall I dig under her skin for her bones? Then when I die, I cannot enter her body to be born again.[7]

This attitude sounds far removed from the European outlook. It is interesting to discover that such values existed in Europe and had to be dismantled before large scale commercial exploitation of natural resources could take place. A belief that persisted from Greek and Roman writers until well into the eighteenth century was that the Earth produced minerals and metals within her reproductive system. Several ancient writers warned against mining the depths of Mother Earth and Pliny in his *Natural History* (circa AD 78) argues that earthquakes are an expression of anger at the violation of the Earth:

We trace out all the veins of the Earth and yet are astonished that it should occasionally cleave asunder or tremble: as though these signs could be any other than expressions of the indignation felt by our sacred parent![8]

Pliny also suggests that the results of mining are of dubious benefit to humanity: that gold has contributed to human corruption and avarice, while iron has led to robbery and warfare.

Such restraints against the exploitation of resources were still active until the Renaissance, but they were dismantled as commercial mining gathered momentum during the fifteenth century. In an allegory which was published in Germany in 1495, this conflict between respect for the Earth and the commercial interests of mining was dramatized in the form of a vision. A hermit falls asleep and dreams that he witnesses a confrontation between a miner and Mother Earth – "noble and freeborn, clad in a green robe, who walked like a person rather mature in years". Her clothing is torn and her body has been pierced. She is accompanied by several gods who accuse the miner of murder.[9] Bacchus complains that his vines have been fed into furnaces; Ceres states that her fields have been devastated by pollution; and Pluto says he cannot reside in his own kingdom because of all the hammering going on. In his own defence, the miner argues that Earth "who takes the name of mother and proclaims her love for mankind" in reality conceals metals in her inward parts in such a way that she fulfils the role of a step-mother rather than a true parent. Since gold turned into money is the surest way to create wealth, the mining of metal will help the poor, will decorate the churches as an act of piety, and will advance culture through the building of schools and the payment of teachers. Thus in these early stages of capitalism, nature's image is changing from something to be respected (the mother) into a source of wealth that needs to be forced into revealing things (the selfish step-parent).

A further significant step in this process is to be found in the work of Francis Bacon (1561–1626) who is celebrated as the founding father of scientific research, and the innovator of inductive reasoning. Bacon was also a politician: Lord Chancellor of England under James I, who had written a book on witchcraft (*Daemonologie*, 1597). In 1603, the first year of his English reign, James passed a law condemning all practitioners of witchcraft to death.

In her account of the scientific revolution, *The Death of Nature*, Carolyn Merchant suggests a striking congruity between the scientist's interest in nature, and the state's interest in witchcraft.[10] Bacon proposes an experimental methodology for investigating nature, using language which is starkly sexual in its metaphors and suggestive of a witch-finder in its techniques. Advocating a scientific method which will be experimentally verifiable, displaying consistent results when repeated, Bacon writes: "For you have but to follow, and as it were hound Nature in her wanderings, and you will be able to lead her and drive her to the same place again." Describing the investigation of nature in the laboratory, Bacon invokes the language of the torture chamber: "I mean in this great plea to examine Nature herself and the arts upon interrogatories. For like as a man's disposition is never well known till he be crossed, nor Proteus ever changed shape till he was straitened and held fast, so Nature exhibits herself more clearly under the trials and vexations of art (mechanical devices) than when left to herself."

The contrast between Bacon's attitude towards nature in his scientific project, and the image of the Earth as an elderly lady who has been assaulted (as reported in the hermit's dream of 120 years earlier) is striking: one can see the characterization of nature as a woman changing to allow research and exploitation to take place. Bacon even described matter itself as a "wanton harlot": "Matter is not devoid of an appetite and an inclination to dissolve the world and fall back into the old Chaos." Nature must be "bound into service" and "made a slave", "put in constraint" and "molded by the mechanical arts".

The clockwork universe

Feminist historians have argued that the persecution of witches in the sixteenth and seventeenth centuries can be understood only as part of the revolution in values that took place as capitalism superseded the feudal system.[11] A new order came to predominate in the seventeenth century which can be characterized as male, mathematical and mercantile. Male surgeons replaced female midwives in Britain, and women who practised medicine were likely to find themselves stigmatized as witches. Mercantile capitalism viewed nature as a resource to be exploited, as can be seen from the example of mining. And scientific research produced a mathematical model of the universe, which superseded the organic analogies of the animate Earth.

Copernicus (1543) attributed to the Earth a daily rotation on its axis, as well as an annual orbit around the sun, because this was a more mathematically concise system; Kepler, equipped with the data of Tycho Brahe, conceived a universe which was constructed from harmonious geometric relations. Galileo (1638) succeeded in establishing mathematical principles which predicted the paths of both heavenly bodies and projectiles on Earth. For Galileo, "Natural philosophy is written in that great book which lies before our eyes – I mean the Universe – but we cannot understand it if we do not first learn the language and grasp the symbols in which it is written. The book of nature is written in the language of mathematics."[12]

Galileo followed Kepler in differentiating between primary and secondary qualities, but in a more pronounced form. Primary qualities included number, motion, and weight. These qualities were absolute, objective, and above all could be quantified mathematically. Secondary qualities included colour, taste, odour, and these qualities were relative, and above all subjective. These three scientists, through their observation of the heavens and their measurements of phenomena on Earth, drafted a mathematical model of the universe which amounted to a re-definition of nature. The real world lay outside of men and women – it was mathematics and astronomy and the study of motion. Objective nature was whatever could be measured.

Finally, Sir Isaac Newton stood on the shoulders of the scientists who had preceded him and reduced the major phenomena of the universe to a single mathematical law – the universal law of gravitation (1687). "The fundamental task of all science is to explain all phenomena in terms of matter and motion", proclaimed Newton.[13] This view of the universe not only altered the status of man, it also made God slightly redundant. In the mediaeval world picture, God had been the "ultimate good" and man's purpose was to know God and to love Him. In Newton's system, God had been relegated to the role of laboratory technician whose job was to prevent the fixed stars falling out of the heavens, and to mend the defects of the celestial clockwork. Newton's favourite proof of the existence of God was the concentric rotation of all the planets in the same plane. The "ultimate good" was now the cosmic order of masses in motion, and man's role was to applaud the workings of this clockwork universe. It was an account of nature which could still (just about) be celebrated in religious terms, as in the Anglican hymn:

> What though in solemn *silence* all
> Move round the *dark* terrestrial ball?
> What though no *real* voice nor sound
> Within their radiant orbs be found?
> In reason's ear they all rejoice
> And utter forth a glorious voice,
> Forever singing as they shine
> 'The hand that made us is divine'.[14]

Presented this way, one can see a certain majesty in the scientific revolution's scheme of order. But in many ways, it's a profoundly alienating vision. Whatever we experience in the way of emotions or sensations is dismissed by science as subjective. What is real is simply the movement of molecules.

From mathematical to pastoral nature

An interesting way of visualizing the changing models of nature is to look at the history of gardens. Changes in the philosophy of nature are reflected in the human versions of nature that kings and gardeners construct. Perhaps the best gardening metaphor for the mathematical model of nature bequeathed to us by the seventeenth century is to be found in the garden constructed at Versailles for Louis XIV. From 1661 until 1690, under the direction of the king's gardener Le Nôtre, thousands of workers struggled to construct a magnificent formal garden out of a swamp. (There was a high death rate due to malaria.) Versailles became the capital of France and Louis XIV's statement of his absolute power.

Thousands of fully grown forest trees were planted in elaborate designs, and when half of them died, they were simply planted again. The domination and control of nature were celebrated in terms of formal design: not a single

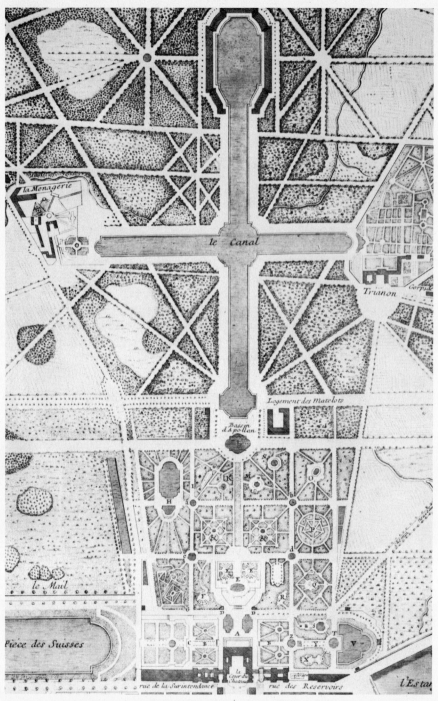

Pierre le Pautre, detail from a plan of 1714 published in *Les Plans, coupes, profils et elevations de la chapelle du chasteau royal de Versailles* (reproduced by permission of the Syndics of the Cambridge University Library).

flower was allowed to wilt, and at Versailles whole beds were re-planted in the course of a day. The power of the king over his subjects, the control of nature by man, are echoed in the way the Château dominates the garden: together they form a statement of mathematical order and of sovereignty.

One hundred years later, a reaction against this account of nature as absolute order was sweeping across Europe. One can glimpse this at Versailles: hidden away in a corner of the king's estate is a fairy tale village constructed for Marie Antoinette, wife of Louis XVI. It is called Petit Hameau (Small Hamlet) and includes a mill, a farmhouse, and a dairy grouped around a lake. This village formed a backdrop for acting out a fantasy version of rural life: the queen and her servants would dress up as milkmaids and shepherds and frolic on the grass. This was the degenerate end of a view of nature known as the pastoral.

Pastoral poetry forms a tradition within Western culture that stretches back two thousand years to the verses of Virgil and Theocritus. The pastoral celebrated nature as restorative: an antidote to the sophistication and cynicism that characterised life at court. Later poets (such as Milton in *Lycidas*) used the pastoral as a way of attacking corruption in the church and in moral matters. They were not attempting a realistic description of country life, but engaging in a moral critique. The pastoral can be viewed as the seed from which sprang the Romantic view of nature: nature as embodying moral values. Nature as a corrective to man's pride. But by the eighteenth century, the pastoral had become the self-indulgence of Marie Antoinette's village, the amorous make-believe to be found in the paintings of Boucher and Fragonard: a fantasy of rural life untouched by work or poverty. Handel's pastoral opera, *Acis and Galatea*, has a chorus of ecstatic peasants singing:

> O the pleasures of the plains
> Happy nymphs and happy swains
> Harmless, merry, free, and gay
> Dance and sport the hours away.[15]

The pastoral was degenerating into a mood of Arcadian escapism at the very moment when farmers and land-owners were beginning to systematically 'improve' the performance of nature. The advent of seed-drilling, the introduction of new crops and a more scientific approach to the breeding of livestock were all features of early eighteenth-century agriculture. If the pastoral were to retain any plausibility, it had to find a way of engaging with the realities of management and work within the eighteenth-century landscape.

The English artist Thomas Gainsborough toned down the artificiality of pastoral painting and drew on more realistic accounts of country life that were being painted in Holland by Hobemma and van Ruysdael. But to reconcile the detailed materialism of the Dutch style with the idyllic fantasies of French artists, such as Watteau and Poussin, produced tensions when it came to representing work.

Thomas Gainsborough, *Landscape with a Woodcutter Courting a Milkmaid*
(reproduced by kind permission of the Marquess of Tavistock, and the
Trustees of the Bedford Estates).

In his study of the rural poor in English painting, *The Dark Side of the
Landscape*, John Barrell has shown how Gainsborough tried to reconcile the
work of the countryside with pastoral pleasure.[16] In *Landscape With A
Woodcutter Courting A Milkmaid*, a ploughman guiding his team across a field
merges with the background, while in the foreground the attractive young
couple (taking time off from their milking and woodcutting) are engaged in
their courtship. Gainsborough has harmonized the countryside at work and
at play. Some of his other paintings display a striking contrast between the
male and female characters. *Peasants Going to Market: Early Morning*
depicts men who look convincingly ragged and tired (like real peasants in the
early morning), while the women seem inexplicably dignified and well-dressed.
Gainsborough is caught up in the contradictions of country life that we still
live with: the countryside is presented as a backdrop for carefree relaxation,
but it must also deliver the agricultural goods.

In perhaps his best-known image of the English countryside – his portrait
of Mr and Mrs Andrews in the National Gallery – Gainsborough integrates

Thomas Gainsborough, *Mr and Mrs Andrews* (reproduced by courtesy of the Trustees, the National Gallery, London).

three very different versions of nature. In the background is untouched wilderness; to the left of the couple is the picturesque area, landscaped to please the eye; and on the right of the picture is a field of abundant corn, painted with loving detail to show that Mr Andrews was an agricultural improver, and that by drilling his corn he was obtaining a bumper harvest. The expressions on the faces of Mr and Mrs Andrews (a look of "intense ownership") remind us of another change that was transforming the landscape of Britain: enclosures.

A sense of the picturesque came to dominate the land-owning classes' view of their property. Instead of nature's being marshalled into a grand design (as at Versailles), human mastery over it had become subtler, more softly modulated, more naturalized. From the increased wealth generated by commerce and industry, a rural retreat could be constructed as an escape from the smoke of the city. Through technical breakthroughs in drainage and earth-moving, it was now possible to re-arrange the landscape to suit the owner's taste. As Raymond Williams describes it, the English landscape garden was a *tour de force* of stage design, but fully realised in physical terms: "a rural landscape emptied of rural labour and of labourers; a sylvan and watery prospect with a hundred analogies in neo-pastoral painting and poetry, from which the facts of production had been banished".[17]

A classic example of this form of gardening can be seen at Stourhead in Wiltshire, designed by the owner, Henry Hoare II, a banker, and constructed under his direction between 1740 and 1780. This garden cannot be seen from the house, and was planned rather as the setting for walks and rides within a picturesque and literary landscape. Hoare had the valley to the west of his house dammed, creating a lovely series of lakes amidst wooded slopes. And a sequence of temples and mythological sites suggest literary associations to the visitor making his way round this carefully constructed "natural" landscape.

If nature was being fenced off for aesthetic enjoyment, then what had happened to the agricultural workers? One answer can be found in Oliver Goldsmith's poem *The Deserted Village* (1769) which describes the clearance of the villagers, so that a landscaped garden can be constructed:

One only master grasps the whole domain

with the consequence that

The man of wealth and pride
Takes up a space that many poor supplied
Space for his lakes, his park's extended bounds
Space for his horses, equipage, and hounds.

Two natures: industrial and Romantic

By the end of the eighteenth century, the industrial revolution was drawing the poor out of the countryside and propelling them towards the factories and cities. Around the year 1840, Britain became the first country where the urban population exceeded the rural population. The violence of this social transformation had a strong effect on the values that attached to the country and the city. More and more, the countryside was perceived as that which had been lost, that which had been left behind. From this perspective of social upheaval, William Wordsworth articulated the moral values that resided in nature. A nature that was indissolubly linked to his childhood, growing up in Cumbria, and to a sense of the fragility of that vision:

> And I have felt
> A presence that disturbs me with the joy
> Of elevated thoughts; a sense sublime
> Of something far more deeply interfused;
> Whose dwelling is the light of setting suns,
> And the round ocean and the living air,
> And the blue sky, and in the mind of man:
> A motion and a spirit, that impels
> All thinking things, all objects of all thought,
> And rolls through all things. Therefore am I still
> A lover of the meadows and the woods
> And mountains; and of all that we behold
> From this green earth; of all the mighty world
> Of eye and ear – both what they half create,
> And what perceive; well pleased to recognise
> In nature and the language of the sense
> The anchor of my purest thoughts, the nurse,
> The guide, the guardian of my heart and soul
> Of all my moral being.[18]

Nature had become something we communed with, something we yearned for. Emphatically separate from human society and everyday life. It's worth pausing to discuss what it was that alienated and mystified nature into this fleeting, transcendental force. The short answer would be industrialisation and city life.

As the centre of gravity of society was drawn towards the vast and fast-growing cities of the nineteenth century, the countryside and nature were felt to be repositories of a lost way of life, and of values that were in danger of extinction. A rural community was felt to be organic and rooted in the soil; in contrast to urban life with its anonymity and sense of constant change. Obviously the idea persists that closeness to the flow of nature – the smell of damp earth, the cyclical growth and decay we associate with the seasons – is somehow a more 'natural' way of life. And yet the industrial world is simply based upon another model of nature which enables the resources of the world to be analysed and broken down and re-shaped by human hands. Francis Bacon had pointed this out in 1623, when he described three forms of nature:

She is either free and follows her ordinary course of development as in the heavens, in the animal and vegetable creation, and in the general way of the universe; or she is driven out of her ordinary course by the perseverness, insolence, and forwardness of matter, as in the case of monsters; or lastly, she is put in constraint, molded, and made as it were new by art and the hand of man, as in things artificial.[19]

Today, a common meaning for nature is simply an escape from the "man"-made world: an Eden which we look forward to re-gaining at the weekend, or on holiday. It is a common experience to walk through a beautiful, wooded landscape, and then to notice a line of electricity pylons and reflect how human industry has 'spoilt' nature once again. It should be remembered that the electricity pylons, the very notion of an electric current as a controlled flow of electrons, are based upon an analysis and re-arrangement of nature. The most prestigious science magazine in Britain today is entitled *Nature*, since nature remains the starting point of all scientific and technological activity.

In a sense, the scientific version of nature gave rise to the Romantic version: to supply what was missing. Since sensations and emotions had been banished by science from nature, dismissed as subjective, the Romantics expressed their alienation by constructing an account of nature that overflowed with emotions and moral feelings. They tried to replace the precision of science with a sense of mystery and infinity.

(The Romantic attitude towards science was well expressed at a dinner party in December 1817, where Keats proposed the toast, "Confusion to the memory of Newton!" And when Wordsworth insisted upon an explanation before he drank, Keats replied "Because he destroyed the beauty of the rainbow by reducing it to a prism.")[20]

As the poets of the nineteenth century turned their eyes from the squalor and alienation of the industrial city, and looked hopefully towards 'unspoilt' nature, it is worth remembering that they were merely looking from one account of nature to another.

Survival of the fittest

One of the most spectacular examples of changing accounts of nature reflecting changes in society can be seen in nineteenth-century England. Since the discoveries of Newton, the character of nature had been (to use Raymond Williams' image) that of a constitutional lawyer: interpreting and classifying examples, making predictions out of precedents. "What was classically observed was a fixed state, or fixed laws, or fixed laws of motion. The laws of nature were indeed constitutional, but unlike real constitutions they had no effective history. What changed this emphasis was of course the evidence and the idea of evolution: natural forms had not only a constitution but also a history."[21] At the beginning of the nineteenth century, the role of nature shifted towards the selective breeder, improving the species through natural selection.

Perhaps the first significant figure in this process was Thomas Malthus, whose *Essay on Population* (1798) argued that the population was growing too quickly, and that finite resources of food would soon be exhausted.[22] The replenishing cycles of nature would be wiped out by uncontrollable growth. Therefore such catastrophes as plague and famine were not aberrations, but an important part of nature's design. There were not enough places at nature's dinner table to feed everyone, suggested Malthus. Famine provided the pruning shears to thin out nature's garden. Harvest failures were examples of nature's red pencil auditing the book of life. The dinner table, the gardening shears, the audit book: all these images of a prudent, commercial society were used to legitimise a profound callousness towards 'natural' catastrophes. Naturalists had begun to analyse and classify the animal kingdom; soon this spirit of detached observation gave rise to a similar look at human society.

Darwin's account of evolution, *The Origin of Species* (1859), was remarkable for the controversy it provoked about whether mankind was part of nature, as well as for the more ferocious character that was attributed to nature as a consequence of Darwinism.[23] When Bishop Wilberforce, a mathematician and ambitious prelate, debated with Thomas Huxley, one of the first scientists to declare his support for Darwin's theories, Wilberforce asked Huxley, "was it through his grandfather or his grandmother that he claimed his descent from a monkey?" The activities of Biblical fundamentalists in the United States who, in the 1970s, have tried to replace the theory of evolution with the Book of Genesis, may remind us that this question is not yet dead.

But though the theory of evolution has become a part of scientific orthodoxy, perhaps Darwin's work has had a still more profound (if unintended) effect on ideas of human society. Fifty years after Wordsworth invoked nature as "the guardian of my heart and soul", another poet laureate, Alfred Lord Tennyson, described nature as "red in tooth and claw" and invoked a world of unlimited savagery and strife:

> A monster then, a dream,
> A discord. Dragons of the prime
> Which tear each other in the slime.[24]

This world of course bears striking similarities to mid-nineteenth-century Britain: this was the peak of industrialization, the working class had been drawn into the factory towns and a ruthless exploitation of workers and of natural resources was in progress, recorded by Engels and by Charles Dickens. On an international scale, Britain and the other world powers were

Sir Edwin Landseer, *Man Proposes, God Disposes* (reproduced by kind permission of Royal Holloway College, London).

busily seizing and exploiting less advanced countries and cultures. Suddenly, Darwin's imagery and argument provided the perfect rationalisation of this process. "Survival of the fittest" had initially referred to "the best adapted to the environment". Out of a series of chance mutations, those creatures that successfully occupied an ecological niche flourished, while others died out. Soon 'the fittest' came to mean the most powerful, the most ruthless.

Artists as well as poets dramatically revised the personality of nature, moving from the pastoral to the violent. The most celebrated nature painter of mid-nineteenth-century Britain, Sir Edwin Landseer, painted landscapes of natural carnage. *The Swannery Invaded by Eagles* depicts a group of snow-white swans being torn to shreds by sea eagles. *Man Proposes, God Disposes* features two thuggish polar bears munching the bones of some hapless polar explorers. And an early work by Landseer, *The Cat's Paw*, illustrates an ingenious act of animal sadism: a monkey seizes a cat's paw and forces it against a red-hot stove. Many of his contemporaries were disturbed by Landseer's skill at representing animal savagery, and it is difficult to look through the body of his work without feeling that these images of animal violence are really statements about society.

The way in which Darwin's argument influenced accounts of society (referred to as Social Darwinism)[25] is obvious from such everyday figures of speech as the rat race, the pecking order, the law of the jungle. As soon as Darwin's theory had become popular currency, historians and social scientists saw history in a new light. Karl Marx, at work on *Das Kapital*, acknowledged that Darwin's theory of natural selection could serve "as a basis in the natural sciences for the class struggle in history". Marx wrote to Darwin, offering to dedicate *Das Kapital* to the great scientist. But his collaborator, Friedrich Engels, warned that: "The distinguishing feature of human society is production and when the means of production are socially produced, the categories taken from the animal kingdom are inapplicable." Similarly, the anarchist Kropotkin pointed out that Darwin's argument concerned competition *between* species, whereas Social Darwinism seeks to justify competition *within* a species: between members of the human race.

And yet there has been no shortage of analogies from Darwin to society because the idea seems so suited to competition for jobs on the labour market, competition between companies for a share of the market, competition between nations for territory and power. In short, the salient features of capitalist, industrially advanced, and politically rivalrous powers. John D. Rockefeller argued that you cannot grow a perfect example of the American Beauty rose without first debudding it of minor blooms; in the same way, a company cannot achieve maturity without eliminating serious competitors – in other words, establishing a monopoly.

The German commander von Moltke (who provided the military muscle behind Bismarck's unification of Germany) saw war as the supreme example of the Darwinian struggle for existence, where the armies of competing nations clashed and only the strongest survived. On a global level, it could even be argued that European nations had a moral duty to compete with 'inferior' races, in order to raise the quality of the species.

Managing the domain

By the end of the nineteenth century, Europe was in command of Africa and vast areas of Asia. The perspective on nature shifted from the state of all-out war that characterised the 'Darwinian' outlook, to the need for efficient management. Having conquered foreign territories, the task was to set up an efficient government. Having eliminated your industrial competitors, the job was now to consolidate your market. Having collected all the animals and plants, it was now necessary to catalogue them.

Major museum collections were systematised to give order to the specimens that had been slaughtered around the world. In Britain, Lord Rothschild despatched expeditions to remote corners of the globe to collect butterflies, and the Rothschild Collection ultimately amounted to two-and-a-half million set butterflies and moths, meticulously labelled and mounted, 300,000 birds, 144 giant tortoises, and 200,000 birds' eggs.[26] Some naturalists have speculated that this spirit of insatiable collecting and cataloguing is a consequence of the Protestant ethic at work in the animal kingdom, since the major European collections were to be found in Britain, Germany, and Sweden; in Catholic countries this tendency towards monumental collections of nature hardly existed.

The Royal Botanic Gardens at Kew housed a similar collection of plants uprooted from all over the world. Research and selective breeding of plants took place at Kew; important crops were made more 'productive' and then transplanted to other locations within the British Empire. In particular, the rubber plant was successfully 'kidnapped' from its home in the Amazon

Basin, improved upon, and then transplanted to Malaya where it became the basis of a multi-million-pound industry.[27]

Even in municipal parks and gardens, it became common to incorporate ducks and trees transported from other continents. Henry Hoare's garden at Stourhead was transformed by importing exotic rhododendra from North America and the Himalayas, so today the garden has a very different character from its original one. Nature as a global resource was being gathered together and efficiently cultivated.

Simultaneously, the warfare between town and country was mellowing into the idea that these two states could be harmoniously integrated. A key work was Ebenezer Howard's book *Tomorrow: A Peaceful Path to Real Reform* (1898) which became the blueprint for the garden city movement. Howard's diagram for his "slumless, smokeless city" was a series of concentric circles with six boulevards radiating out from the centre, where the major civic buildings were located. Industrial, commercial, and residential zones were sited in different parts of the city and broken up by 'green belts'. The outermost circle was an agricultural zone which would supply the city with food, and also each house was to have an extensive garden to produce vegetables for the family. Through this scheme, Howard believed the most alienating effects of industrial capitalism could be transformed: instead of being divided by competition, the city would flourish through a spirit of co-operation. The polar opposites of town and country would be superseded by "a third alternative, in which all the advantages of the most energetic and active town life, with all the beauty and delight of the country, may be secured in perfect combination".

The first garden cities were built at Letchworth, Hampstead, and Welwyn; these formed the starting point for a peculiarly British form of Utopian planning. Another generation of New Towns were built after the Second World War, and in the 1960s Milton Keynes was started as the largest (and probably last) city in this line of thinking. To some extent, the vision of transforming society was reduced to a series of town planning procedures; the Housing and Town Planning Acts of the twentieth century incorporated many of Ebenezer Howard's ideas, and the project of reconciling country and city became a bureaucratic system.[28]

Primary qualities take over

The Renaissance astronomer Rhäticus suggested that if you can measure something, then you have some control over it. This was an interesting metaphysical notion. The French mathematician Laplace took things one stage further when he postulated: "If a superhuman intelligence were acquainted with the position and motion of every atom in the universe, then this intelligence could predict the entire future of the universe." It was assumed that this mathematical hypothesis could never be put to the test. But with the growth of computer technology, and more and more sophisticated ways of gathering information, encoding it, processing it, and disseminating it, one can see Laplace's remark as more than just abstract speculation. The more the world is analysed and quantified, the more sophisticated the intelligence gathering and control systems become. (In some countries, there are now laws which give citizens access to the information which is being stored about them; this is democracy struggling to survive in an electronic age.)

What has all this to do with nature? It could be argued that a world characterised by an explosion of information and information processing systems is the culmination of the project that began with the scientific revolution: to de-code the book of nature as mathematical information. The essence of computer technology is that information can only be fed in once it has been encoded in digital form. Any information that cannot be mathematically encoded, cannot be considered. The Romantics rejected the world view that resulted from Newton's discoveries because they felt that the most important things had been left out – feelings and moral values.

Today, the Newtonian world view is certainly in control of the planet, and this growth of information systems has amounted to a new industrial revolution. The first industrial revolution analysed nature in order to imitate it mechanically, and improve its efficiency. Steam engines were far more powerful and efficient than water wheels or windmills, the old ways of deriving power from sources of energy. The new industrial revolution is electronic rather than mechanical, and is concerned with encoding information and simulating aspects of the human nervous system. Machines that can count. Machines that can calculate. Machines that can read and write, that can talk and listen, that can see and analyse what they see, that can recognise voices and faces. The more complex the imitation of the human brain, and the more it is linked to mechanical systems which can act on that information, the closer we come to a world which exists as a vast simulation of so-called reality. (Several science fiction novels and films have picked up on this idea by devising plots in which the hero somehow gets sucked inside a computer simulation of the world, and then he has to find a way of escaping back into the real world.)

Children growing up today take for granted the programming and use of computers. They play games on electronic toys which simulate battles between space invaders. They are used to recording audio-visual information on video and digital systems. And they grow up with ordinary TV sets giving them access to vast amounts of information – weather reports, rates of exchange, news flashes – which are updated every minute. For kids growing up in technically advanced countries, this electronic universe is already second nature to them.

The ways in which these systems amount to forms of political control are complex and quite subtle. When Rhäticus suggested that measurement is a form of control, he did not necessarily mean straightforward coercion. The earliest Greek scientist, Thales of Miletus, caused a sensation in 585 BC when his astronomical measurements enabled him to predict an eclipse of the sun. His mathematical knowledge was a primitive form of power. Knowledge itself brings control.

In more down to earth terms, one can also notice how this electronic revolution is automating thousands of jobs out of existence. Word processors will replace many secretaries and clerks. And it is now common in any high street to programme some digits into an electronic cash dispenser, and be handed some bank notes by a machine. When computers are linked to mechanical systems, the result is the science of robotics which enables human workers to be mechanically replaced. The jobs that do remain can be paced and subjected to close surveillance. Systems for monitoring movements and for recording and analysing phone conversations, telex messages, radio signals, are all in the forefront of this electronic revolution.

An electronic imitation of the world might seem to be the antithesis of nature – but it derives from the way in which Galileo and Newton re-interpreted nature as a mathematical system. Today could be characterised as the cybernetic era of nature – nature as bits of information. This is the culmination of the doctrine of primary qualities. And we are left to seek some consolation in the world of secondary qualities – smell, sound, sensations, emotions – and to flirt with some of the personalities which nature has discarded.

The end?

Town planner, ruthless predator, selective breeder, "guardian of my moral being", constitutional lawyer, agricultural improver, celestial clock-maker, "wanton harlot", loving mother: all these characters have been attributed to nature in the past five hundred years. Which one do we live with today? The short answer would be all of them. When we watch a TV programme about nature, we are likely to be startled by close-ups of creatures tearing each other to shreds. If we switch to a gardening programme, we may see middle-aged men with moustaches speaking tenderly about a young plant they are nurturing, and touching its leaves as gently as a baby. If we visit the natural history museum, we can study nature stuffed and mounted and painstakingly catalogued. If we take a holiday in the Lake District or the Alps we may be overwhelmed by the grandeur of nature. If our holiday is interrupted by an earthquake, we are reminded of nature's monstrous, de-structive forces. If we simply go for a walk in a park, we are likely to glimpse a version of either the mathematical order of Versailles, or of the carefully constructed landscape at Stourhead. And if we shop for natural shampoo or for herbal remedies, perhaps we are shopping for a reminder of the days when the Earth was alive.

These are all encounters with nature which take place every day in modern Britain. But beneath these examples there are larger issues at stake. Earlier I mentioned the experience of walking in the country and being annoyed by the sight of electricity pylons, and said this could be seen as one account of nature intersecting another. More dramatically, in the women's peace camp at Greenham Common, the representatives of one version of nature have confronted the representatives of another. These women have spoken about their concern for the future of the Earth. Some say that as mothers they have a special responsibility for the world their children will inherit, and they object to a piece of the English countryside being hollowed out to house nuclear missiles. And Greenham is just one small but visible part of a system which transforms the Earth's resources into instruments of death. Yet these weapons cannot be dismissed as 'unnatural': they are the most striking examples of human analysis and command of nature. The very structure of the atom has been "put in constraint, molded, and made new by art and the hand of man" – as Francis Bacon put it.

Implicit in this account of the scientific model of nature has been a critique of technology which I believe reflects a widespread anxiety about the future of our world. It is this anxiety which underlies the conflict at Greenham Common. Reading accounts of technology from the nineteenth and early twentieth centuries, what is striking is the optimism they display about technology's ability to fulfil human needs. This optimism has been undermined by the way in which technology seems to go on creating new, unanticipated needs. It has become an end in itself. Traditionally, the scientific revolution has been represented as the triumph of reason. The simplicity of turning on an electric light, the ease of flying by jet from one continent to another, the supply of drinking water for vast cities – all these benefits of technology have enhanced our lives and today are taken for granted, but each year our awareness grows of the price to be paid for this knowledge. A sense of ecological crisis, and a fear of nuclear annihilation are currently the two most obvious examples of this price.

Inevitably, this alienation has turned most of us into Romantics – at least, in our private lives. We look to nature as a repository of values and we experience the open countryside as a release from the pressures of work and everyday life. But a glance at the history of Greenham Common shows that this 'unspoilt' piece of nature has served many different ends: it used to be a stretch of common ground for grazing cattle and sheep. For centuries this area was supported by the wool trade, and in the early years of the industrial revolution, textile machinery was driven by the local streams. Poorly paid agricultural workers rioted here in 1830 and were hunted down by Grenadier Guards: one was hanged and many were transported. During the Second World War, the United States Air Force arrived. From Greenham, scores of gliders took off as part of the invasion of Europe on 6 June 1944. In 1951, despite widespread local opposition, the Air Ministry acquired most of

Greenham Common for conversion into a permanent air base. And in 1981 the government announced that cruise missiles would be deployed here.

Perhaps this land and everything that has happened on it, is a simple way of seeing that nature is what we make it. There is no force out there called 'Nature' which justifies our actions, and which may hopefully come and save us from this polluted and alienated world. Nature is our own creation, and may yet prove to be our own destruction. If we wish to live in peace with each other, and with this planet, it is our responsibility to write that into the next chapter of the book of nature.

Notes

This essay developed from a documentary film which I produced and directed for Channel 4, entitled *A History of Nature*. The script was written by myself and Dr Robert M. Young, and I would like to acknowledge that many of the ideas and examples in this essay are indebted to my close collaboration with Bob Young, though the responsibility for this article is mine alone. I would also like to thank John Barrell, Maureen McNeil, Richard Patterson, and Conrad Atkinson for the help they gave on this film, with special thanks to Carolyn Merchant and Jane Cousins. And I would like to acknowledge the influence of Raymond Williams' work, which goes beyond the references here cited.

1 For a fuller account of the meanings of the word 'Nature', see Raymond Williams, *Keywords* (London: Fontana, 1976), pp. 184–9.

2 See Raymond Williams, *Problems in Materialism and Culture* (London: Verso, 1980), 'Ideas of Nature', pp. 67–86.

3 This use of the term 'man' as a generic to cover the whole human race is deliberate. Whenever this term occurs in this article, it is referring to a specifically male-dominated order – from God the father, at the summit of mediaeval theology, through the court and society of the Sun King, to the manufacture and deployment of nuclear weapons.

4 See Arthur Lovejoy, *The Great Chain of Being* (Harvard University Press, 1936).

5 See Carolyn Merchant, *The Death of Nature* (London: Wildwood, 1982), 'Organic Society and Utopia', pp. 69–99.

6 Athanasius Kircher, *Mundus Subterraneus* (Amsterdam, 1665). See Marjorie Nicolson, *Mountain Gloom and Mountain Glory* (New York: Cornell University Press, 1959), pp. 168–73.

7 Smohalla of the Columbia Basin Tribe in early 19th century, quoted in Alfonso Ortiz and Margaret Ortiz, eds., *To Carry Forth the Vine* (New York: Columbia University Press, 1978).

8 Pliny, *Natural History*, trans. J. Bostock and H. T. Riley (London: Bohn, 1858), vol. 6, Bk 33, ch. 1.

9 Niavis, *Judicum Jovis* (Leipzig, n.d.). For a fuller account of this allegory, see Frank Dawson Adams, *The Birth and Development of the Geological Sciences* (New York: Dover, 1938), pp. 171–5.

10 See Merchant, *The Death of Nature*, ch. 7: 'Dominion over Nature', pp. 164–91.

11 See Merchant, *The Death of Nature*; Susan Griffin, *Woman and Nature* (New York: Harper and Row, 1978); Barbara Ehrenreich and Deirdre English, *Witches, Midwives, and Nurses* (New York: Feminist Press, 1972).

12 See Edwin Arthur Burtt, *The Metaphysical Foundations of Modern Physical Science* (London: Kegan Paul, 1925), ch. 3: 'Galileo', pp. 61–95.

13 Burtt, 'Metaphysical Foundations', pp. 228–99.

14 'The Spacious Firmament on High', hymn written by Joseph Addison (London: *The Spectator*, 1712).

15 *Acis and Galatea*, music by George Frideric Handel, libretto by Alexander Pope and John Gay (London, 1718).

16 John Barrell, *The Dark Side of the Landscape: The Rural Poor in English Painting 1730–1840* (Cambridge University Press, 1980).

17 Raymond Williams, *The Country and the City* (London: Chatto & Windus, 1973), ch. 12: 'Pleasing Prospects'.

18 William Wordsworth, *Lines Composed a few Miles above Tintern Abbey. July 13, 1798*.

19 Bacon, 'De Dignitate et Augmentis Scientarum' in *Works*, ed. J. Spedding, R. Ellis, D. Heath (London: Longmans, 1870), vol. 4, p. 296.

20 For an account of the poetic response to Newton, see Marjorie Nicolson, *Newton Demands the Muse* (Princeton University Press, 1946); M. H. Abrams, *The Mirror and the Lamp* (Oxford University Press, 1953), pp. 298–335.

21 Williams, *Problems in Materialism and Culture*, 'Ideas of Nature', p. 73.

22 See R. M. Young, 'Malthus and the Evolutionists: The Common Context of Biological and Social Theory', *Past and Present*, 43 (1969), pp. 109–45.

23 See R. M. Young, 'The Historiographic and Ideological Contexts of the Nineteenth Century Debates on Man's Place in Nature', in M. Teich and R. Young (eds.), *Changing Perspectives in the History of Science* (London: Heinemann, 1973), pp. 344–438.

24 Alfred Tennyson, *In Memoriam* (published 1850), stanza lvii.

25 See Williams, *Problems in Materialism and Culture*, 'Social Darwinism', pp. 86–103; R. M. Young 'The Human Limits of Nature', in J. Benthall (ed.), *The Limits of Human Nature* (London: Allen Lane, 1973).

26 See Miriam Rothschild, *Dear Lord Rothschild* (London: Hutchinson, 1983).

27 Lucile Brockway, *Science and Colonial Expansion: The Role of the British Royal Botanic Gardens* (London: Academic Press, 1979), pp. 141–65.

28 See Ian Tod and Michael Wheeler, *Utopia* (London: Orbis, 1978), pp. 119–26.

2

The societal conception of space*

ROBERT SACK

We have concentrated on the essential symbolic structures of modes of thought that are potentially a part of the intellectual capacities of individuals in all societies. These modes are social in the sense that individuals producing them are influenced by society and the organizations or communities to which they belong. These social conditions undoubtedly affect the contents of the modes and the degree to which they are separate and distinct. There are, however, realms of activities that are predominantly group, collective or social, and in which the individuals act in the name of the group. Nations, cities, armies, families and scientific, religious and artistic organisations are examples of collective or social relationships or facts. These facts are linked to space in two related ways. First, the social organisations and the individuals within them are 'in' space and their interactions have spatial manifestations. Thus we have families in cities, and cities in regions containing other cities, and so on. Analysing these relationships has been the traditional concern of social geography and has resulted in such theories as central place, land use, and the gravity and potential models.

Second, social organisations are often territorial, a fact largely overlooked by all but some political geographers. Territoriality here does not mean the location and extension in space of a social organisation or of its members. Rather it means the assertion by an organisation, or an individual in the name of the organisation, that an area of geographic space is under its influence or control. Whereas all members of social organisations occupy space, not all social organisations make such territorial assertions. The social enforcement (and institutionalisation) of such assertions – in the form of property rights, political territories or territories of corporations and institutions – provide the context necessary for social facts to exhibit the first type of spatial properties. The forms such territorial structures take and the functions they provide depend on the nature of particular political economies.

* Source: Robert David Sack, *Conceptions of Space in Social Thought: A Geographic Perspective* (London: Macmillan, 1980), pp. 167–93.

Editors' note: the original includes a substantial number of detailed footnotes. We have reduced these to a small number of references. Readers interested in the sources used by Sack should refer to the original book.

Both the spatial relations and territoriality of social facts involve distinct conceptions of space which we will refer to as *societal conceptions*. We will explore the societal conceptions of space associated with the political–economic structures of society as a whole.

Mixtures of modes

The greatest differences among conceptions of space at the level of societies as a whole are found when we classify social systems in a general evolutionary schema from primitive societies to civilisations, with some civilisations evolving into modern nation states. (We will not consider the various transitional stages between the primitive and civilised such as chiefdoms.)

There are two primary properties to the societal conception of space which apply especially to the level of the political–economic structures and which most clearly illustrate the differences in views of space that are associated with the differences between primitive and civilised. The first property is the conception that a people have regarding the relationship between their society and its geographic place. As with other things, societies occupy space. The first property refers to a people's conception of this relationship. Societies tend to forge strong ties to the places they occupy and to justify these ties through social organisations and procedures. Different societies conceive of these ties to place differently. In some primitive societies the social order is not thought of as possessing a continuous extension in physical space. Rather, the society is anchored to the earth's surface in very special locations such as holy places, sources of water and traditional camp sites. The intervening areas, although known to the members, may be unimportant to them in a territorial sense. In such cases the territorial boundaries would tend to be vague. For other societies, the social order may be conceived of as extensive over space where the boundaries may be more or less clearly defined and may become territorial. In civilised societies, parts of the society are seen as possessing continuous extension, but what parts and how clearly their boundaries are defined differ from one type to another.

The second property of the societal concept of space is the knowledge and attitude that a people have regarding other peoples and places. In such cases we are more interested in the elaborateness of the spatial viewpoint than in the specific details or content of the knowledge.[...] There are, for example, primitive societies that have virtually no knowledge of other places or peoples except their own. Their view is extremely ethnocentric and space is literally the place or territory they occupy. Beyond the place, the idea of space does not apply. Other primitive societies may have a slightly more elaborate view of other people and places. They may be conscious of where other people border them, but, except for this, they may be unaware of their neighbours' territories and of what extends beyond. In civilizations, with the possible exception of feudal societies, we generally find a more articulated view of the territories beyond one's own, a view that conforms to a projective or Euclidean space.[...]

The two properties are roughly interrelated with regard to their degree of sophistication and these degrees are roughly related to the division of society into primitive and civilised. In societies which do rely heavily on unsophisticated–fused views, the social order is often mythologised and linked to place through a mythical–magical mode. Similarly, the conception of the space surrounding the society is seen through this mode and in fact becomes submerged within it. In civilisations (and all civilisations have elaborate sophisticated modes of thought), both the anchoring of society to place and the relationship of the society to other places can be viewed in a sophisticated pattern. But these societies incorporate unsophisticated modes as well. Which pattern predominates depends on the particular structure of the society and on its relationship to other societies.

The primitive*

[...]The kind of society which is referred to as primitive belongs before the rise of ancient civilisations some 7–8000 years ago. There are, of course, no such societies left to observe. Our informed views about them come from archaeological reconstructions and from anthropological field work in preliterate societies which have comparable technologies to those unearthed by archaeology. Even though the 'primitive' societies studied by modern anthropologists have changed in the last 8000 years and have been in contact with more developed societies, their technologies and social structures are radically different from modern ones and closer to what we know of the older societies from archaeological remains. Therefore, with proper precautions, we can use contemporary ethnographic data as evidence for a plausible characterisation of the earlier societies. We will concentrate especially on those features which set the primitive apart from civilisations and which most clearly illustrate their societal view of space.

Primitive groups are less complex than civilisations. They have less division of labour, internal specialisation, fewer numbers and smaller territories. But among them there are different orders of complexity, ranging from the bands and clans of the hunters and gatherers through the more complex tribal societies to perhaps the highest forms of primitive social organisation, the tribal confederation, such as the Iroquois of north-eastern United States or the Maori of New Zealand. In all primitive groups, the family is the basic unit and often the next higher social unit is the band.

For the most part, the hunting and gathering bands are non-sedentary. A

---

* Editors note: in the original, Sack devotes some space to a discussion of the term 'primitive'. We have omitted this as it is peripheral to our focus on the relationships between space and society. We do, however, share Sack's view that the term 'primitive' should not be regarded as pejorative.

band may have a different ecological habitat for each season. Their numbers never approach the size of even a modest town. Of the extant hunters and gatherers, perhaps the Eskimos have the largest villages, numbering in places of good hunting, several hundred inhabitants. But most villages are much smaller. Their primitive technology makes for little specialisation and division of labour beyond that of age and sex. Their nomadic existence makes the family unit the essential core of those societies. Links beyond the nuclear family are established through conceptions of kinships.

The tribe is more complex than the band. The term covers a range of societies occupying different habitats, and having different economics and population sizes. In the tribe, as in the band, the family is the core unit. However, the kinship links in tribes are much more precise and extensive than in bands. Tribal settlements may contain only a single primary family, but the average size of the villages of non-intensive agricultural tribal groups is approximately 1–200. In intensively cultivated areas, tribal settlements could be as large as 1500.

Tribes have a more segmented social order than do clans and bands, but not until we come to civilisations do there exist true economic *classes*. In tribes, the division of labour is still predominantly based on age and sex. Community life in the tribe is family oriented. The community seems to provide its members with an intimate and enveloping sense of belonging. The 'naturalness' of this unit is one of the most striking aspects of primitive societies. The sense of a unified community, as Diamond put it, 'spring[s] from common origins, [is composed] of reciprocating persons and grow[s] from within'. The sense of community is enhanced by the tendency in primitive societies to use the family as an analogy for society and its relationships with the world. This analogical extension of the family contributes to the general primitive view of the unity of nature. It creates a personalism which extends to nature and which underlies, and is perhaps the most distinctive element of, primitive thought and behaviour.

The intimate link between person and community does not stifle individuality and personal expression. In fact, according to many observers, there is greater allowance for individual expressions in primitive societies than in civilised ones.

How individuals see themselves in relationship to the community is difficult to determine. Most observers agree that the bond between individual and society among the primitives is unusual and unusually close. This impression has led observers to believe that in some respects the individuals do not conceptually separate themselves from the group; that is, the conception of individual and group is prelogical. As Lévy-Bruhl has said of primitive societies, "the individual as such, scarcely enters into the representations of primitives. For them he only really exists insofar as he participates in his group".[1][...]

It is their wariness about and avoidance of abstractions that more than anything else are the essential characteristics of non-mythical–magical primitive thought. Many of their so-called philosophical statements are in fact admonitions against abstractions. We in the West, for instance, are expected to love, and to believe that love is a good thing. The primitives tell us not to love love, but to love specific people, and to demonstrate such love rather than pronounce it.

The primitives' tendency to shy away from abstractions may pertain especially to their thoughts about society because the primitive social order does not present the need for social abstractions. Primitive society lacks such powerful elements as economic inequality and class conflict which would create social differences that only abstract social philosophy or revolution would help to reconcile. There are no legal, political, administrative institutions, organisations or apparatus of state apart from and above the people. Primitive society is participatory. Conflicts in society are not directed against institutions or corporate entities, but against specific individuals. Their arguments are not abstracted into political theories which offer alternate conceptions of social order. When an individual finds a social rule or norm to be an insurmountable obstacle, he may break the rule, or leave the group to form a new group wherein the rules of conduct may very well be a copy of ones of the society from which he left. The rule is not seen as an obstacle for the attainment of a particular goal and may never be an obstacle again. Because conflicts are personal and not abstract, such societies do not produce political and social alienation. Opponents are people, not institutions or group entities. Opposition is not between 'we the people' and 'they the society'. There simply are no revolutions in primitive society. In Diamond's terms,

The primitive stands at the center of a synthetic holistic universe of concrete activities, disinterested in the causal nexus between them, for only consistent crises stimulate interest in the causal analysis of society. It is the pathological disharmony of social parts that compels us minutely to isolate one from another, and inquire into their reciprocal effects.[2]

The lack of need for abstractions about society at the primitive level explains the scanty and problematic evidence of its occurrence. It lends weight to the view that in primitive society the individual and society are not thought of as very distinct elements, because there is no need for such abstractions in primitive societies.

The organic relationship between individual and society is recapitulated in the relationship between society and milieu. As the individual is not alienated from society, society is not conceived of as independent of the place which it occupies nor are individuals alienated from the land. A constant and intimate knowledge of place enveloped by a mythical view of the land fuses the society to place. Place is often inhabited by the spirits of the ancestors and a specific place may have been given to a people by their gods. In Australia each totemic group is associated with a place from which the totemic ancestor is supposed to have emerged. When a person dies their spirit returns to the place of their totemic origin.[...]

Physiognomically arresting landscape forms are often the ones incorporated into the myths, helping to anchor the society to place. According to Penobscot Indian lore, much of the landscape is a result of the peregrinations of the mythical personage Gluskabe. The Penobscot river came to be when he killed a monster frog, Gluskabe's snowshoe tracks are still impressed on the rocks near Mila, Maine. A twenty-five foot long rock near Castine is his overturned canoe, the rocks leading from it are his footprints, and Kineo mountain is his overturned cooking pot. A place on the earth in many creation myths was given to a people specifically by the gods. The Pawnee, for instance, believe that they were guided from within the earth to their present place by Mother Corn, and the Keresan Pueblo Indians believe that they were led by Iyatiku, their mother, from the centre of the earth to a place on the earth's surface called Shipap.

Belief in the inhabitation of the land by the spirits of ancestors and in the mythical bestowal of the land to the people have occasioned a powerful communal sense of ownership and use. To have access to the land one must be a member of the society, which means partaking in the spiritual history of the group. For example, in Bakongo tradition,

the ownership of the soil is collective, but this concept is very complex. It is the clan or family which owns the soil but the clan or family is not composed only of the living, but also, and primarily, of the dead; that is the *Bakulu*. The *Bakulu* are not all the dead of the clan; they are only its righteous ancestors, those who are leading a successful life in their villages under the earth. The members of the clan who do not uphold the laws of the clan...are excluded from their society. It is the *Bakulu* who have acquired the clan's domain with its forests and rivers, its ponds and its springs; it is they who have been buried in this land. They continue to rule the land. They often return to their springs and rivers and ponds. The wild beasts of the bush and the forest are their goats, the birds are their poultry. It is they who 'give' the edible caterpillars of the trees, the fish of the rivers, the wine of the palm trees, the crops of the field. The members of the clan who are living on the soil can cultivate, harvest, hunt, fish; they make use of the ancestral domain, but it is the dead who remain its guardians. The clan and the soil it occupies constitute an indivisible thing, and the whole is under the rule of the *Bakulu*. It follows that the total alienation of the land or a part of it is something contrary to Bakongo mentality.[3]

Society and place were so closely interrelated that for the primitive to indulge in speculation about the society elsewhere or about the society having a different spatial configuration, would be like severing the roots from a plant. It could be of no value. But such intellectual contrivances are precisely what social planning and theory require. Statements like 'what if the social order were altered so that land were held differently'; 'what if the village were redesigned, placing this here rather than there, making that rectangular rather than circular, so that certain goals will be more easily attained', are the basis of a conceptual approach to society and place. They involve the conceptual separation and recombination of social activities or substances from space – a separation which underlies all forms of social theory. This separation and attempted recombination of space and society are absent in the primitive world. The place and the people are conceptually fused. The society derives meaning from place, the place is defined in terms of social relationships, and the individuals in the society are not alienated from the land.

Civilisation

[...] While the reasons for that transition from primitive to civilisation are unclear, it is well established that in all of the changes there was a replacement of the classless society by a class society. Civilisations have economic classes. The factor of economic class is extremely important for it leads to different and often contradictory and antagonistic views of social order.

No longer was society seen by all of its members in the same light. Rather, it became a multifaceted and ambiguous concept. In order for civilisations to cohere, such different views would have to be reconciled. The mechanism concurrent with the formation of civilisation which served to reconcile these problems was the formation of the state, whether it be the archaic state, the feudal state, the oriental state or the modern state.

The state stands as an institution above the people encompassing the entire society. Its parts seem not to be equivalent to the citizenry. Unlike primitive society, the government of the state and its officials and power are distinct from the people and their powers. The government has the power to coerce its citizens. The state, as Engels explains,

is a product of society at a particular stage of development; it is the admission that this society has involved itself in insoluble self contradiction and is cleft into irreconcilable antagonisms which it is powerless to exorcise. But in order that these antagonisms, classes with conflicting interests, shall not consume themselves and society in fruitless struggle, a power, apparently standing above society, has become necessary to moderate the conflict and keep it within the bounds of 'order;' and this power, arisen out of society but placing itself above it and increasingly alienating itself from it, is the state.[4]

To make this power more accessible, visible or 'real', the state is endowed with the most basic attribute of objects – location and extension in space. In civilisation, political power of the state is areal or territorial. The state is reified by placing it in space. Territorialisation of authority provides an open-ended assertion, and, if successful, exertion of control. By expressing power territorially there does not need to be a complete specification of the objects, events and relationships, which are subject to the authority of the state. Anything, both known and unknown, can fall under its authority if located within its territory. This open-ended means of asserting control is essential for the political activities of civilisations because it involves, almost by definition, the confrontation of novel and the unforeseen events; and such an unspecified domain can be claimed only through areal or territorial authority. Whatever else a state may be or do, it is territorial.

The linking of society to place is more of a conscious effort in civilisations than it is in primitive societies and its function in the former is more clearly to reify a power and authority which, because of vastness and complexity of civilisation, is not clear and self-evident. Consciously moulding society into a territory tends to place more emphasis on the *territorial definition of society* than on the *social definition of territory*. The former has meant that social relationships are determined by location in a territory primarily and not by prior social connections, whereas the latter has meant that the use of an area or territory depends first and foremost on belonging to a group (the determination of which is essentially non-territorial). In civilisation, a person's domicile frequently determines the person's membership in social organisations. Each location may be part of several overlapping or hierarchical jurisdictions so that being a resident of a place often means being part of several communities.

While a more territorial definition of power is a characteristic of civilisation, the degree to which all members of a society are aware of this varies tremendously. No civilisation has attempted to make its citizenry more aware of the central authority and the territorial extent of that authority than the modern nation state. Yet, many states in the twentieth century contain tribal and peasant cultures which are only minimally affected by modern conceptions of political territoriality and land use.

In addition to containing tribal cultures, a large proportion of the population of developing states contains peasant societies whose conceptions of place often combine elements of both primitive and modern. The peasants' conception of place focuses on the village and its surroundings. Their relationship to the land is extremely close, personal and often mystical. Yet, their livelihood depends in part on contact with the city and the government, and from such contacts, some 'modern' conceptions have been incorporated within the more traditional and inward-looking views of peasant communities.

A far greater mixture of civilised and primitive existed in the ancient states. Civilisations arose slowly and only partially replaced primitive societies. Often, the two societies lived side by side, or primitive forms of community existed on the local and rural levels within the territorial limits of the state. Much of the detail of the original transformations from primitive to civilised is lost and we have only a general idea of the process in some places. In these cases, there is also evidence of the accompanying changes in the association of society to place.

In ancient Egypt, for example, the original tribal or clan territorial units were called *spats* or *nomes*. As Egypt progressed from a tribal to a centralised empire, the *spat* remained a basic territorial unit but its relationship to the people changed. These units were no longer the demarcation of tribal holdings, but instead became an administrative area or province of the empire; and the position of leaders of the *spat* changed from the chiefs of the older tribal communities to administrators or governors of the province.

For Greece and especially Attica, there exists documentation of the end of the transition from tribal to civilised society. At the beginning of the historical record, the Athenians still had vestiges of tribal structure. In the Heroic age they were composed of four tribes which were settled in separate territories. The tribes were composed of phratries and clans and the government of the Athenians was a tribal council or *Boulē*.

As the society of Athens became more complex, its inhabitants became more intermixed geographically but more clearly demarcated into classes, with the political and economic power being concentrated in the hands of the few in the upper class. The social territorial changes were expressed by the introduction of such institutions as the *Naukrariai*. The *Naukrariai*, established some time before Solon, were "small territorial districts, twelve to each tribe", which were to provide and equip a warship and horsemen for the city of Athens. According to Engels, this institution was one of the earliest recorded examples of the shift to a territorial definition of society. It attacked the older tribal form of association in two ways. First, "it created a public force which was now no longer simply identical with the whole body of the armed people; secondly, for the first time it divided the people for public purposes, not by group of kinship, but by *common place of residence*".

Cleisthenes established a new constitution which completely ignored the older tribal territories. The new order was based on a territorial organisation of society. "Not the people, but the territory was now divided: the inhabitants became a mere political appendage of the territory." Yet, to maintain sentiment to place the new territorial definition of society had to appeal to the older social definition of territory. In this regard, to keep sentiment to place alive, the new territory was a 'local' tribe. But the territory or local tribe was an artificial geographic unit for the convenience of the state. It was not an area which traditionally belonged to a group of related people as in the older tribes which were abolished. The basis of the territorial organisation were approximately 100 districts called *demes*. Residents of each *deme* elected a president, treasurer and judges. To again tap the older tribal attachments to place, the *demes* had their own temple and patron or divinity. The *demes* were organised into thirty *trittys* of approximately equal population, and from these *trittys* were formed ten 'local tribes'. The local tribe raised arms and men for the military and elected fifty representatives to the Athenian council which was composed of 500 representatives; fifty for each of the ten local tribes.

Such changes in the relationships between society and place can be seen in other areas of the world. Civilisation, with its emphasis on the territorial definition of society, makes social order closely bound to place. Community membership is often decided by domicile, and territorial defence (unlike the vagueness of the concept in tribal societies) becomes a primary obligation of the state. An attack on territory is a challenge to the state's order and authority.

It is the social complexity, inequality and the need for control of one group by another which make the territorial definition of society essential in civilisation. Paradoxically, it is the same forces which make the fusion of society to place in civilisation far more tentative and unstable than in the primitive world. On the one hand, territories constrain flows and movements of goods. On the other hand, separating activities in places creates specialisation which in turn increases the demand for trade and circulation. Activities which are readily located or contained within the territorial boundaries come to be thought of at the societal level as place specific, territorial or, in general, as spatial activities, whereas the flows and movements which are not readily contained come to be thought of as the less spatial or non-spatial activities.

Moreover, on the territorial or spatial side there develop degrees of 'spatiality'. Civilisations have several kinds of territorial units existing simultaneously. Modern nation states, for example, are divided into several sub-jurisdictions and these in turn may be divided into lower order districts creating areal hierarchies of territorial units whose actions may often be uncoordinated and in conflict with one another. A hierarchy of jurisdictions, or even a national territory divided into a single level of lower units, can create circumstances which are thought to be more or less spatial. A policy which is formulated at the national level may have predictable spatial consequences at that level, but perhaps not at the lower level of the geographical scale. Hence the spatial consequences of an action become foreseen at only one artificial scale. From the viewpoint of those who want to know the spatial consequences at the lower levels, or at all levels, the policy is not specific enough, and hence 'less' spatial than was desired.

The kinds of activities which come to be seen as less or non-spatial and the degree to which these conceptions are held by the citizens depend on the dynamics of the societies and on the ways in which the societies are anchored to place. In each society, the 'spatial', 'non-spatial' distinctions can appear at various levels of abstraction, from the material level of institutions and economics to the level of political and social philosophy.

Western Feudalism

The specific fusion of society to place in Western Feudalism differed from the fusion in the Graeco–Roman civilisations and in our own. Feudalism began as the absorption of a disintegrating Roman Empire by the Germanic tribes; it was an absorption and accommodation of a civilised view of territory by a primitive one. The Roman institutions of *precarium*, which was a dependent form of land holding, and the *patrocinium*, which was the offering of one's services for protection, were to form the basis of the fiefs and the vassalages of the feudal system. A peasant swore allegiance to a lord, and promised to meet such obligations as paying taxes on crops and rendering specific services to the lord, and in return the lord was to provide protection for the peasant. The entire structure rested on the labour of the peasant who did not own the land on which he lived. The peasant's lord owned the land but only in degree, for he too received the land with obligations from a noble of a higher order, and so on, up the social scale until the highest level – the king. In theory, there was to be no land without a sovereign. *Nulle terre sans seigneur*.

The peasants were bound to the earth, *glebae adscripti*. As long as the peasants fulfilled their part of the obligation they could not in theory be removed from the land. However, the peasant could not choose to leave the land either. While a lord could not sell a peasant as one could a slave, he could exchange land with another lord and, as a result of the exchange, would come a new lord for the peasant. For the peasants, social obligations and relations at the societal level could be determined by domicile. The survival of some communal village land and peasant holding from pre-feudal times did not appreciably weaken the peasant's bondage to the land.

Towns existed in the interstices of the manorial system. While towns were usually modest in size, they were important elements of the medieval economy and formed a fusion of place and society which was coexistent but in many respects contrary to the manorial system. A different form of law pertained within the territorial area of the city. "City air", as the German saying went, "makes man free". These special laws originated from the need of the merchants to have a permanent place of trade in the towns so that they could set up their wares and stay during inclement seasons. The inhabitants of these special places needed the right to travel, and those who sought the merchants' goods needed to be free to come and go. Such liberties of movement were granted, and they were paradoxically place specific. "Freedom became the legal status of the bourgeoisie, so much so that", according to Pirenne, "it was no longer a personal privilege only but a territorial one, inherent in urban soil just as serfdom was in manorial soil. In order to obtain it, it was enough to have resided for a year and a day within the walls of the town."[5]

Craft guilds formed another important aspect of the territorial authority of towns. These guilds established a craft monopoly within the towns. They participated in the government of the community and it was difficult for a person to engage in manufacture or become a labourer in towns without being a member of that community's guilds. These guilds were for the most part not areally interrelated (mercantile guilds were). Each tended to operate within the bounds of the city.

Both city and manor formed different, yet contemporaneous, territorial social organisations. The territorial unit of the city expanded in influence and helped transform the manorial system to commercial agriculture. The domination of the city occurred in part because of the increased importance of such activities as the flow of goods, people and money which were not containable within the existing territorial boundaries. Such flows which were not containable by the manorial system later became the basis of the less or non-spatial activities in industrialised societies. Even in the Middle Ages, such activities were not accorded equal status with property in land as a real thing. Property in land was, and is still, called *real estate*. The other is called *liquid assets*.

Although the coexistence of such territorial units as cities and manorial systems engendered processes which were incapable of territorial confinement or expression, the primary 'a-spatial' factor at the societal level was the idea of the Christian community. Indeed, the Church, as an organisation, was spatial. It was territorially administered and appeared on the ground in the form of churches and holy places. These territorial aspects often conflicted with the territorial units of the secular society and also with the more ethereal concerns of the Church. But from the societal view, the primary opposition which the Church presented to the feudal order was the concept of an 'a-spatial' community of Christians. This community or heavenly city transcended terrestrial communities and boundaries. Unlike the 'non-spatial' aspects of society in the contemporary world, the Christian community of the church transcendent was associated with the fixed and the eternal, while the earthly cities were short-lived and changing.

The idea of a transcendent Christian community pervaded Christendom, had an enormous impact on feudal society, and contributed to the later conflicts between church and state, in which the power of the state triumphed. It was the rise of capitalism, however, which most fundamentally contributed to territorial re-organisation within the state.

Capitalism

With the rise of merchant and then industrial capital, the means of production became concentrated in the hands of capitalists and spatially concentrated in workshops and manufactories wherein the labourer worked for wages under the supervision of management. Work thus became separated from the home and territorial control became specialised to include work places (under the control of industry and supported by government through property laws etc.) and political territories. The new economic order needed large and mobile labour pools and 'free' trade with a reliable and efficient transportation infrastructure. All of these factors altered the older territorial fusion of society to place. Coordination of economic functions was achieved by shifting the basic fusion of society and place to the larger geographic scale of the absolute state and then to the modern nation state. This left the cities with modest powers, as one of several territorial units in an areal hierarchy of state territorial organisations. Inhabitants of a city were not only citizens of that city, but also citizens of higher and lower political administrative units in which their residences were located.

As we well know, the anchoring of society to place in the nation state, with its hierarchy of territorial units, has not prevented conflicts between types and levels of territorial organisations, nor contained what are thought to be the less spatial processes. If anything, the opposite is the case. As the efforts to contain and anchor the system become more complex and self-conscious, as more levels of administration are created, the activities and interrelations that involve the spatial and non-spatial distinctions increase enormously. Because of the areal hierarchy of the political administrative system, the spatial manifestations of decisions are fragmented. Decisions may be directed to only one level of the hierarchy and yet affect all levels in unforeseen ways. Jurisdictional conflicts arise concerning the territorial units and the actions of one has unforeseen consequences on the others.

Antitheses between 'spatial' and 'non-spatial' facts or activities can be seen in the economic realm as well. In capitalist societies, parcels of land, clearly demarcated in place, are privately owned and held for purposes of speculation. The land may have been purchased because of its potential value, based on what might happen on or near it. Although the land contains substances, such as soil and vegetation, and perhaps even social factors like low-income families, the value of the land to the speculator may be determined solely by the future activities that could occur near or on it. A new highway may be constructed nearby which would increase the value of the property as a commercial site. Even after the highway is built, the owner of the land may not sell or build until the price is right. In such cases, the economic system makes us think of land as though it were empty, void of substances that have value, and of substances as though they were a-spatial entities existing abstractly somewhere but not materially on the land. Only under profitable circumstances do particular places and substances combine, and then only until a more profitable arrangement appears to disassociate and recombine them with other places and things.[...]

The maintenance of social order in twentieth century society cannot rely entirely on the sophisticated but tentative links which social science and scientific planning provide. In actual practice, the fusion of society and place is accomplished through the combination of sophisticated and unsophisticated models. The state uses the sophisticated models of social science, it uses the arts and it taps the unsophisticated models to anchor behaviour to place. The United States has a 'heartland', it has national monuments, shrines and 'holy' places such as the Capitol, the Lincoln Memorial, Grant's Tomb; it has a mythologised past in the rugged frontiersmen and the 'noble savage'. Attachment to nation states in fact may be one of the clearest expressions of mythical–magical consciousness of place in the twentieth century. As Yi-Fu

Tuan points out, sacred space tends to be a locus of power; it is clearly demarcated and set apart; it is supposed to be complete; and it demands the ultimate sacrifice for its defence.[6] All of these characteristics apply to the modern nation state. Each state, though, has different conceptual separations of people and place and depends upon different mixtures of the sophisticated and unsophisticated to recombine them. Understanding these differences is the spatial perspective to social dynamics. What marks the modern conception of societal space from the primitive is the range of mixtures available to the modern society, and the fact that the modern, unlike the primitive, has access to more specialised sophisticated views. With this access and the complexities of social life which engendered such modes, comes a high degree of uncertainty, detachment and scepticism about the significant relationships between social order and geographic area.

Notes

1 L. Levy-Bruhl, *The 'Soul' of the Primitive* (Chicago: Henry Regnery, 1971), p. 185.

2 S. Diamond, *In Search of the Primitive* (New Brunswick, N.J.: Transaction Books, 1974), p. 192.

3 Fr Van Wing, *Etudes Bakongo*, deuxième édition (Desclée de Brouwer, 1959), pp. 93–4.

4 F. Engels, *The Origin of the Family, Private Property and the State* (New York: International Publishers, 1972), p. 176.

5 H. Pirenne, *Economic and Social History of Medieval Europe* (New York: Harcourt Brace and World, 1937), p. 51.

6 Yi-Fu Tuan, *Space and Place* (Minneapolis, University of Minnesota Press, 1977).

PART 2

Introduction
Analysis: aspects of the geography of society

JOHN ALLEN

The message of the introduction to this book was straightforward: geography matters to all of the disciplines in the social sciences. Social processes necessarily take place in geographical space and in some relation to nature, and this carries a series of implications for all explanations of social activity. Neither sociologists nor economists, it should be said, are opposed to the argument that space and nature have a part to play in social explanation; they are not unaware that social activity occurs in space or that nature impinges upon social action. It is not part of our argument that the social science disciplines today are simply blind to these features of the social world; rather, it is that they have failed to conceive the extent to which space and nature are integral to an understanding of social activity and social change.

Space is not simply a surface upon which changes, say, in the structure of the British economy are played out. Changes within the structure of economic production involve the geographical reorganisation of labour and capital; and, in turn, the changing geography of economic activity affects the shape and composition of the workforce and throws up new cultural patterns and political configurations. Changes in the technology of production are central to this process of geographical and social reorganization – changes which involve a rearrangement and control of natural forces to achieve profitable and competitive conditions of production. Gold captured this view of nature in his chapter when he spoke of the electronic imitation of the human brain, microprocessor technology, or 'nature as bits of information' as he refers to it, as the latest objective quantifiable arrangement of nature. This way of approaching the relationship between society, space and nature, this kind of geography, alters the manner in which economic, political and cultural change is conceived.

The implications of this view, however, are not merely conceptual, they also raise methodological questions. Conceptual and methodological issues are woven together to produce a particular kind of geographical approach. The methods of analysis and synthesis employed in the following two sections do not in themselves produce a new kind of geography. The terms are familiar within the discipline of geography. What is different about their usage here is the conceptual framework that governs their application.

Briefly sketched, the two modes of approach, analysis and synthesis, are best understood as complementary methods that enable us to understand the geographical organization of society, both at a detailed level and at a richer, more comprehensive level. The process of analysis involves the selection and isolation of a particular aspect of society, the consideration in detail of its various features and the relationships that hold between them, including their geographical characteristics, at the expense of other aspects of society to which they are related. The value of this type of exercise rests upon a number of pragmatic grounds. First, we cannot hope to take up every aspect of society that has some bearing upon, say, the changing structure of Britain's economy, and simultaneously study all aspects with the same analytical intensity. In a social world structured by a series of complex, changing interrelationships, the process of analysis allows us to hold certain relationships apart and to concentrate in detail upon each aspect in turn. This internal focus is the nub of analysis; it allows an investigation *in depth* of one aspect of a broader picture. Second, some kind of division and separation of the different aspects of the geography of the social world is necessary in order to enable us to address particular social issues and problems. A geography of housing, of welfare, of employment, is the practical development of the need to outline handlable issues and formulate specific answers to specific questions. Which issues are defined as manageable, which questions are raised is not pre-given, and invariably rests upon the institutionalized sub-divisions that hold sway within a discipline at any one point in time.

Sketched in this manner there is, as indicated earlier, nothing that is particularly new either to geographers or to other social scientists about the task analysis is expected to perform. What is distinctive about the selection of readings in the following section, however, is that the analyses of the three topics – cultural forms, urban economic activity, and the process of international law – are framed in terms of how their geographical organization affects the very way in which they work.

Two of the three topics – cultural forms and international law – do not readily spring to mind as geographical issues. Indeed, we have deliberately chosen them to illustrate the importance of geography to all aspects of social activity. None of the three authors, in fact, are professional geographers, yet each acknowledges the importance of geography as an integral part of social explanation. It is an integral component, first, because the variation in social conditions within and between countries affects the manner in which social processes operate in different localities; and second, because this pattern of geographical unevenness is continually in a process of transformation as part of the general dynamic of social change.

Clarke's analysis of cultural forms in the UK draws the two points out sharply. In Chapter 3 he explores the geographical unevenness of local cultures, their unique characteristics, and challenges the thesis of a mass culture that pervades all areas and regions of the UK. From this standpoint he shows how the cultural differences between places have had a profound effect upon the shape and absorption of new cultural configurations. New cultural forms, he argues, are not simply mapped onto local cultures: the relationship between the two is one of negotiation and resistance, 'traces' of earlier local cultures affect the form of new cultural strands. In turn, he argues that the geography of culture cannot be understood without reference to structural changes in the occupational division of labour. Changes in the structure of capitalist production which have reworked the social and gender composition of the workforce also, he points out, involve a geographical shift of capital and labour. The movement of both industry and workers to new locations has led to the disruption of old cultural boundaries and created a new pattern of cultural unevenness and flux. Cultural, economic and spatial change are, he argues, combined in the overall process of social change.

The links between economic, political and cultural processes are built into his analysis, but they are deliberately played down to allow a more intensive focus upon changing cultural traditions. In an analytical sense, Clarke has turned a spotlight upon cultural processes to show how culture itself cannot be fully understood outside of its geographical context.

Ball and Picciotto draw out the same point in relation to urban economic activity and international law respectively. In Chapter 4, Ball demonstrates how the cost of crossing space measured in terms of time, information and money influences and shapes the way in which different types of economic activity, both private and public, use urban space. The barriers of distance may be used, for example, by retailers to gain a monopoly position over a particular clientele by locating at a distance from their competitors' shops. In this sense, space is integral to the activities of retailers and cannot be analysed in isolation from their economic behaviour, nor can their economic behaviour be considered in isolation from spatial opportunities and spatial form. At the level of the city as a whole the spatial form of social processes can produce contradictory results. As the tertiary and quaternary sectors take over the city centre they push up land prices there, and their interest in land and location as a long-term asset pushes them up even more. One result is that low-income housing becomes, given the politics and economics of the production of housing, less and less of a feasible proposition. And yet it is precisely the office-owners of the tertiary and quarternary sectors that demand, close at hand, the nightly army of low-income workers to clean up after the day's business.

In the following chapter, Picciotto addresses the growing incidence of jurisdictional conflicts between major capitalist nation states. In a number of examples ranging from the financial sanctions against Iran to the US embargo policies against the Soviet bloc, Picciotto graphically illustrates how the divergent political interests and economic rivalries between nation states assert themselves as geographical barriers to the international integration of the world economy. To put it another way, the increasing internationalization of capital has geographically outstripped the political process of internationalization. Political territoriality, as Sack pointed out in his article (Chapter 2), is more than the geographical space occupied by a nation state, it is an assertion of power and control over the events and relationships that occur within a defined space. The anchorage of states to place which allows the defence of economic interests within a territorial unit is shown by Picciotto to be at odds with a process of capital that has outgrown state political boundaries. Conflicts over territorial jurisdiction have arisen as the concomitant of the interpenetration of capital investment between the developed capitalist economies. The changing geographical structure of the international economy and the political world order are integral to the form in which the conflicts have arisen and are played out.

One further point about the mode of analysis conducted in these chapters should be mentioned. Each topic has been conceptualized in such a way as to establish its relation to the wider aspects of society which affect its social form and development. The analyses are not simply 'cultural' or 'economic' or 'political', they do not reflect any particular sub-divisions of systematic geography, wider connections are built into the analyses to locate the topic within the larger social order. This is not to deny the particular focus of analysis, it is merely to indicate that the topics remain linked to the broader social context.

In Chapter 4, for example, Ball's analysis of urban economic life does not exhaust all that there is to know about the dynamic forces which shape and structure city life. For the purposes of detailed investigation he selects, or abstracts, three elements – the division of labour, population distribution and the built environment – from, say, the political forces and cultural processes which also shape the fabric of city life. In making such a choice Ball is consciously dividing up and separating off different aspects of the geography not only of the city, but also of the wider social world of which cities are a part. But he is also careful to draw the links, for example, between an urban spatial division of labour and the wider processes of capital accumulation and national state policy. His analysis spills over into social relationships that appear only at the edge of his focus, yet indicate the direction in which changes within cities may be understood in a more holistic fashion. This is the groundwork for the construction of a broader synthesis of which the city would be one element.

In a similar vein Picciotto's analysis of jurisdictional conflicts between capitalist nation states indicates that an issue of international law cannot be adequately comprehended outside of its relation to aspects of international economics and politics. The latter two features are undeveloped in his account, the links are faintly drawn, to enable him to portray the issue of legal conflict in sharp relief. Clarke, in Chapter 3, is also insistent that culture should not be seen as simply a set of local differences in isolation from wider aspects of society. The complex geography of culture that he outlines is not conjured up by a variety of local characteristics; local cultural terms draw their changing shape from a combination of such characteristics and the wider changes that have affected the economic organisation of British society, both at a national and at an international level.

3

'There's no place like...': cultures of difference[1]

JOHN CLARKE

There is a mundane level at which everyone is familiar with a geography of British culture. To visit a museum, art gallery, theatre or cinema is to be confronted by the geography of cultural institutions. Trips to the town or city centre consume time and money. To live outside London involves a recognition of the concentration of cultural resources and institutions of all kinds which divides the 'metropolis' from the 'provinces'. To visit Blackpool and Eastbourne is to encounter the diverse sources of pleasure that are associated with British holidays.

These patterns bear the mark of social processes. Here, the 'civic pride' of nineteenth-century philanthropists who stamped their mark on city centres with galleries and museums, and there the 'economic rationality' of the cinema chains, closing the old suburban 'flea-pits' and developing the city centre multiple-screen complexes. Space and place are essential elements in the patterns of British culture, and one of their most profound effects is to be found in the way they structure cultural diversity. Blackpool and Eastbourne offer different holiday pleasures, not solely because of where they are, but because of the different social groups whose holiday needs they cater for. The bank holidays and wakes weeks of a Northern working class, stretched for fifty weeks a year by labour and poverty, required an intensity of sensation and satisfaction garishly reflected in the distorting mirrors of Blackpool's 'cheap thrills'.[2] The inter-war middle classes of the Southeast, by contrast, lived in a culture defined by the virtues of familiar decency and respectability more adequately serviced by the genteel spaces and tea rooms of the South Coast.

These class cultural differences, and their geographical placing, were well understood in inter-war Britain. Cultural space – the placing of class boundaries – had very sharp demarcations.

Culture and place

Place – the region, the city and the neighbourhood – condenses a whole complex history of economic, social and political processes into a simple cultural image. The persistent opposition of North versus South expresses

54

itself in cultural images (hard *v.* soft; rough *v.* cultured; straight *v.* sly) but such stereotypes are meaningless without attention to the economic, social and political differentiation which are condensed in these parodies. The cultural division refracts the economic and industrial unevenness of Britain, and the concentration of economic power and control in the Southeast. It captures something of the distribution of 'cultural capital' which flows alongside that concentration of economic capital, and it gestures knowingly at the focus of political power that sits just around the corner from the City. It is, of course, a parody. It subsumes class within region, as if there is neither Southern working class nor Northern capitalists. But it is a parody which contains a partial, and uncomfortable, recognition of the geography of economic power and control in Britain.

The demarcation of Britain into distinctive cultural regions owed much to the way capitalism focused its development. The North of England was dominated by the urban centres in which the heavy industrial sectors and textiles were concentrated. The Midlands, the Black Country and Birmingham in particular, were organized around metal and secondary engineering production, experiencing an influx of labour from Scotland and Wales. The Southeast, although dominated by the centre of financial and political power in London, was also becoming the favoured site for the development of the new 'light industries' in engineering, electrical and chemical industries. In the Northeast, Scotland and Wales, the older skilled trades, in sectors such as mining and shipbuilding, were bearing the brunt of the inter-war Depression.

This uneven distribution of industrial sectors explains something of the differential impact of the Depression on British society. The decline of the old 'staple' industries, and the unemployment and pauperization which followed, scarred the face of Northern England, Scotland and Wales in sharp contrast to the self-confident expansion experienced in the Southeast. British culture, the ways in which people experienced, understood and responded to these material conditions, bore the marks of this uneven and unequal economic structure. The different 'cultures' which were present in inter-war Britain reflected these disparate economic conditions through a variety of national, regional and local identities. So, the Welsh working class, shaped by chapel, rugby and Celtic cultural legacies, possessed a culture distinct from that of the Northern working class, based around football and cricket, the regional identities (and rivalries) of Yorkshire and Lancashire, and the 'civic pride' of the mill and steel towns. In other ways, the presence or absence of women as a primary group in the labour-force (textiles versus mining, for example) shaped different traditions of trade unionism, local politics and the nature of 'family life'.

It is impossible to deal with the full interplay of Britain's inter-war economic, social and cultural patterns here. Each 'place' was culturally complex. Scotland contained the diversity of agricultural highlands, the Glasgow working class, Edinburgh's 'cultural' cosmopolitanism, and so on.

Similarly, within any town, different areas and neighbourhoods contained specific cultural patterns from the 'exclusive' residential districts to run-down terraced housing. But what should be clear is that place – nationally, regionally and locally – played a major role in organizing these cultural differences, and was an essential *symbol* in the identification of difference. Geordie, Glaswegian, Mancunian, Cockney: each local identity condensed a whole range of economic, social and political references into a *place*.

But this geography of culture is subject to drift, erosion and dramatic eruptions. The processes which forged the class cultures – and their places – of the inter-war years did not stop, leaving these cultures frozen into a static formation. Viewed from the 1980s, these pictures of Blackpool and Eastbourne become nostalgic snapshots of a different time, a different way of life now transformed by social change. The clear boundaries of social distinction have been washed away. Between then and now stands the Second World War, affluence, new technology, the expansion of the mass media – and recession. What, then, has happened to the old places, boundaries and patterns, and, more pertinently, what has replaced them?

Making sense of post-war Britain has preoccupied a variety of social analysts, and their preoccupations have brought forward a variety of answers. One of the most powerful responses to these changes was the assertion that they had *abolished* difference and diversity and resulted in the creation of a 'mass culture'.

An absence of difference: the mass culture thesis

The mass culture thesis pointed to the massive post-war expansion of the mass media (and television in particular) as the force which had abolished cultural difference. The spread of television, and its penetration into the home, were seen as the means through which a generalized culture was being dispersed and absorbed. And since the commercial drive behind the media was concerned for audience maximization, the programming which sought to create these audiences would be based on the 'lowest common denominator' approach. These three elements – spread, penetration and content – form the core of the mass culture thesis.[3]

The spreading ownership of televisions allowed most of the nation to be linked together as a mass receiver of transmitted culture. This linking of a mass audience to a single message promised the obliteration of intranational cultural differences, and even the disappearance of international cultural differences. The communications revolution promised to make the world 'a global village' – the abolition of geography itself – with American its common language.

Secondly, television penetrated to the 'sacred hearth', the world appeared in the living-room corner. Its visual immediacy, allied to its domestic setting, was understood to provide the culture it presented with especially powerful impact. Studies proliferated to examine the impact on the first 'telly generation', assessing the effects of *Dixon of Dock Green* and others on violence, morals, schoolwork, and nightmares.

Thirdly, and here was the cultural rub, the content of mass culture threatened to extinguish traditional standards. Since most mass cultural theorists were, like most mass culture, American, they spoke of mass culture displacing the values embodied in 'high' and 'popular' culture.

'High' culture – the arts – was prized for its engagement of the emotions and intellect in the quest for knowledge and ultimate human values. 'Popular' culture, while not possessing these same virtues, was at least tolerable in that it demonstrated the vitality and robustness of 'the people'. But mass culture threatened to produce nothing but the 'cultural dupe' – the passive and indiscriminate sponge.

All of this would have been alarming enough as a theory, but there seemed to be plenty of both quantitative and qualitative evidence to support it. The post-war consumer boom seemed to indicate that the new-found affluence would erode the distinctive way of life of the 'old' working class. Consumption and leisure patterns seemed set to converge into a mass culture. Hugh Gaitskell pronounced the obituary of the working class after Labour's 1959 election defeat with a rueful eye on the emerging mass culture of consumerism:

In short, the changing character of labour, full employment, new housing, the new way of life based on the telly, the fridge, the car and the glossy magazines – all have had their effect on our political strength.   (Quoted in Hall *et al.*, 1978, p. 230)

The successful Conservative leader put it more abruptly: "The class war is over" (Macmillan, quoted in Hall *et al.*, 1978, p. 227).

Macmillan's conclusion is the pertinent British variation of mass culture, for what its arrival in Britain promised to eliminate was not merely 'high' and 'popular' culture but the fractious and troubling irritant of class cultures.

From Leeds to Luton: the disappearance of class

Gaitskell's summary of the forces which had eroded Labourism were taken up by cultural analysts concerned with the disappearance not only of working class politics, but also of a distinctive working class culture beneath the rolling waves of affluence and mass culture. Studies such as Richard Hoggart's *The Uses of Literacy* (1959) are ambiguous. They are, as Critcher has argued, "a response to...the argument that the working class had ceased to exist" (1979, p. 16). While they celebrate the distinctiveness of working class culture, they also lament its anticipated disappearance. Hoggart's study offers both a sensitive description of a distinctive way of life, and an attempt to analyse the causes of its destruction – those proceses which are "unbending the mainsprings of action". The studies of this period are diverse in their identification of the main causes of change (affluence, mass culture, social policy in housing, etc.) but each of them offers its testimony to the disappearance of something distinctive. And each of them also recognizes that one of the changes is the remaking of the geography of class.

The working class 'community' dealt with in these studies marked the convergence of geography and culture. Whether in a London housing estate (Young and Willmott's study of Bethnal Green), a Northeast mining village (Dennis *et al.*'s 'Ashton') or Hoggart's Leeds, what these studies celebrated was a specific geographical density of social relationships and a shared and distinctive way of life.[4] The economic, political and cultural changes of post-war Britain were identified as destroying the strengths of these densely supportive local class cultures. In their place, the future offered the affluent, individualized, middle-class culture of mass society.

While these cultural and community studies looked to the old working class estates, towns and villages for the distinctiveness of working class culture, other sociologists were looking elsewhere for the signs of the future. Goldthorpe, Lockwood, Bechofer and Platt, in their mammoth study of *The Affluent Worker* (1969), shook the dust of the North from their feet and set off for the Southeast. There, in the mass production systems and high wage economy of the car industry, the future culture of the working class might be discerned. Consequently, in the 1960s it became Luton (and the Vauxhall-Bedford plants) whose entrails were examined for signs of the new working class.

The signs were ambiguous and have been argued over ever since. The 'affluent worker' was not becoming middle class (the victim of 'embourgeoisification'), but nor was Luton the setting for the continuation of the old culture of working class community. Instead the 'new' working class seemed to show an instrumental or calculative approach to employment, trade unionism and politics. Social life – the culture of these workers – had lost the focus of communal sociability and solidarity of the 'community' and replaced it with a 'privatized' or family-centred lifestyle, in which 'place' played no significant part. The factory and the home – not Luton – were the only places which counted.

The Affluent Worker studies suggested that although class (identified in the conditions of work) was not being abolished, there was a 'convergence' in terms of lifestyle, attitudes and beliefs – a cultural convergence. The privatized affluent working class were adopting a way of life – a culture – closer to that of their middle class counterparts.

A difficult phase: youth and cultural diversity

The post-war outbreak of 'youth culture' seemed to represent further evidence of this 'convergence' of lifestyles. Cultural difference seemed now to be based on age rather than class. 'Teenage culture' appeared to be a classless world of distinctive music, clothes, clubs and behaviour. For those searching anxiously for the portents of mass culture, the signs were hardly encouraging. Youth, like the affluent workers, represented a 'vanguard' – the first generation exposed to classlessness, affluence and the mass media. And the behaviour of this new generation reinforced all the fears of the mass culture pessimists: immorality, declining standards, drugs, violence and an apparent surrender to mindless consumption.

Both T.V. channels now run weekly programmes in which popular records are played to teenagers and judged. While the music is performed, the cameras linger savagely over the faces of the audience. What a bottomless chasm of vacuity they reveal. Huge faces, bloated with cheap confectionery and smeared with chain store make-up, the open, sagging mouths and glazed eyes, the hands mindlessly drumming in time to the music, the broken stiletto heels, the shoddy, stereotyped, 'with-it' clothes: here, apparently, is a portrait of a generation enslaved by a commercial machine. (Johnson, 1964).

The obsessive attention to the 'generation gap' as the basis of youth culture successfully prevented any recognition that youth was far from being classless. What was read as being a youth culture was in fact composed of a diversity of youth subcultures in which the structuring hand of class played a powerful role. What was held up as the image of a classless – mass culture – was its opposite – the re-emergence of distinctive lifestyles heavily determined by class. By the time that skinheads were beating up hippies, it was clear that – for them, at least – the class war was not over.[5]

Between the old and the new

In the end, of course, the secret had to come out: class had not gone away after all. A growing body of research indicated the persistence of poverty, gross inequalities of income and life chances and the stubborn refusal of class to be magically forgotten. In part, the excitement and despair about the arrival of mass culture – and the variety of analyses it generated – had been based on a misunderstanding. The very distinctiveness of the inter-war class cultures had deceived the eyes and minds of the analysts. The end of traditional working class culture was understood as marking the end of the working class. The cultural forms of class were taken as being the reality of class.

Because the distinctive working class communities were no longer visible geographically, and because the cultures were no longer visible socially, the assumption was made that class itself must have disappeared. The error was to assume that the *culture* was inseparable from the economic and social basis of class. Little thought was given to the ways in which the basis of class might be refracted through new and changing cultural expressions and practices. Watching television, taking package tours abroad, having a car changed working class culture but did not mark the end of class as an economic and social division.[6]

This mistaken identification of culture with class was intensified by the distinction between the 'new' and the 'traditional' working class. Instead of dealing with the historical process of how classes and cultures have changed and developed, this distinction reduces history to one abrupt change – from the old to the new. The idea of the 'traditional' stops us thinking about history as a process, and leaves history only as a frozen image to be contrasted with the present. But those cultures now described as 'traditional' were themselves the product of economic and social processes, and the responses of social groups to their circumstances.

This is not to suggest that really nothing has changed since the Second World War: that because 'deep down' the class relations of capitalism are still firmly in place, no attention needs to be paid to superficial changes. At a very abstract level, the fundamental dynamic of labour and capital still obtains. But the studies of class cultures of the 1950s do, however erroneously, point to significant changes in the composition of class and culture.

The economic, occupational and social composition of the working class has been reworked by both the dynamic of capital in its search for new products, new markets and new profits and by policies of the state (in housing and other welfare areas). Although all of these processes are interconnected, it is worth briefly separating out some of their impacts. The last forty years have seen a constant process of capital redeployment in Britain – a search for new forms of profitable investment and a wish to escape those with declining profitability. The expansion of the home market for consumer durables – Gaitskell's cars, fridges and tellys – provided the basis for the post-war expansion of light engineering and the car industry in sites largely away from the 'old' heavy industries of the North. Goldthorpe *et al.*'s mission to Luton to find the 'new' (and affluent) worker caught something of this changing geography of the working class.

As the working class became mobile, so the attachments to familiar places became disconnected. New Towns and new housing estates replaced the old communities. Pulled by the promise of affluent lifestyles, and pushed by the changing geography of employment (with the assistance of government regional policies), the working class set out to establish themselves in new places.

But these changes, in the period of post-war expansion, did not become the permanent basis of new social and cultural patterns. By the 1970s, it was becoming clear that the processes of boom and affluence that had produced the new working class were in decline. British manufacturing was entering a deep recession, and the domestic market was being increasingly serviced by cheaper imported goods. Unemployment – the demon of the Depression which seemed to have been exorcised in the post-war boom – was returning. As closures and redundancies increased, capital once again looked around for new sites for profitable investment.

The rise and fall of the car industry is only one (though perhaps the most spectacular) of the occupational shifts, but others are equally significant for the composition of class. The growth of the 'service sector' (both state and private); the decline of the traditional male 'skilled' working class; the expansion of increasingly proletarianized white-collar occupations in technical and administrative systems have played a part in reorganizing the economic physiognomy of class. They have also had profound effects on the sexual division of waged labour. One aspect of the search for profitability has been the attempt to find sources of 'green' (inexperienced and non-unionized) labour, which has (together with the expansion of state and commercial 'service') fuelled the increase of women's employment. This, too, though is uneven – some sectors of 'traditional' women's work (e.g. textiles and clothing) have suffered a similar fate to that of their male equivalent in the manufacturing sectors.[7]

These changing occupational patterns and their uneasy rhythms of decline and expansion have further reshaped the social geography of Britain. The concentration of decline in the old manufacturing industries has been particularly hard on the industrial-urban patterns of the North, Midlands, Scotland and Wales. The major cities and towns have suffered economically and socially, and within each, the process of decline has been most intense in what has become known as the 'inner city'. Equally acute has been the experience of towns dependent on single industries. The 'rationalization' of pits under the NCB removed the economic basis of whole villages and towns, while British Steel's closures at places such as Corby devastated whole local economies.

The rise of new industries has taken place on new sites: oil-based industries in the East of Scotland, electronics in the Southeast 'sun belt' and small manufacturing operations around the New Towns and the industrial 'parks' on the peripheries of old towns. The once stable connection between urban and industrial patterns in British capitalism has been broken – both people and work are moving out of the big cities.

Once more, the patterns are uneven: nationally, regionally and locally. Scotland experienced simultaneous recession (in shipbuilding and the car industry, for example) and expansion (through the North Sea oil fields). Towns given the benefit of industrial enterprise zones to encourage growth lost jobs from companies outside such zones. As the rhythms of expansion and contraction have grown quicker through the post-war period (and as the depths of contraction have intensified), so the certainty of 'place' has been dismantled. Fluidity and mobility became the cornerstones of economic policy, and the ties of place have been loosened in their wake.

Nor can it be said that mobility itself was in any way a guarantee of a new place in the order of things. Workers who moved from Glasgow to the New Town of Linwood to work in the new Rootes car plant at the end of the 1960s experienced several changes of ownership which presided over a declining labour-force before the plant was finally closed.[8]

The economic basis and social geography of class have undergone profound changes. But class is not solely a matter of production – of the organization of work within capitalism. The dynamics of capitalism also affect cultural patterns through the systems of distribution and consumption. The old cultures of region and locality owed as much to the social patterns of home, family and neighbourhood as they did to the conditions of employment. And the post-war period has seen equally substantial changes at this level.

The residential patterns of towns and cities have been reshaped through the expansion of home-ownership, particularly concentrated in the new peripheral estates and New Towns – moving the more affluent out of the older inner city areas. Increasingly, the inner areas have been left to council redevelopment or to owner-occupation of older housing. The rise of the 'property-owning democracy' has intensified the social division and distinctions between centre and periphery in urban areas.[9]

This movement to the peripheries involves the social reorganization of family life – making private transport an essential rather than luxury item. Individual mobility – the ability to get to work, to the shops, to social and leisure activities – has become a necessity for social mobility. The car, while valued for the independence which it provides, has simultaneously created dependence. When public transport has been steadily reduced almost everywhere, those without access to a car have experienced growing social isolation. For the elderly, the unemployed, the young and a majority of women, the 'privatization' of British society, its increasing home-centredness, is as much an enforced condition as it is a freely chosen way of life.

The pressure towards mobility has been intensified by changes in the system of distribution of consumer goods and services. The extent of concentration and rationalization of capitalist production is well understood, but rather less attention has been paid to the consequences of these changes in distribution. The 'old' working class culture had its characteristic commercial institutions – the local shop, pub, cinema, betting shop, and so on. These once familiar elements of the 'community' have also changed with the dynamics of post-war capitalism. Like the industrial sector, they too have been subject to concentration, rationalization – and disappearance.

Consumption, leisure and pleasure now demand mobility. Local pubs were closed or transformed by the marketing expertise of the 'big five' breweries; local cinemas were closed in favour of the city centre multi-screen complexes; betting shops have been subject to take-overs (and the inevitable 'rationalization') by the big leisure companies. The dense sociability (as well as the backbiting and local gossip) of the corner shop has been displaced by the depersonalized relationships of the super- (or hyper-) market.[10]

The local geography of cultural institutions, and the networks of social relations which they make possible, has changed dramatically (except on *Coronation Street*). Occasionally, the consequences of this wholesale destruction of stable geographical patterns has generated a wish to revive 'community'. Through the 1960s and '70s, both local and central government confronted urban blight, 'pockets of poverty', vandalism and other symptoms of decay with a deep sense of nostalgia for 'community spirit'.

The response was to sponsor a whole variety of 'community' based initiatives and projects aimed at reconstructing local pride, local networks, local action – the social virtues which had somehow got lost in the geographical reconstruction of class. Some of these projects were directed to the new estates (which appeared to lack a sense of identity) and others to the unreconstructed inner-city areas still populated by those too old, too poor to too stubborn to take advantage of the benefits of redevelopment. There, some of the 'old ways' lingered on rubbing uneasily alongside an arrival of 'new ways' – of migrant workers seeking housing which escaped the discrimination of both private and state sectors, and of 'young professionals' in search of areas with 'character'.

Traces, sediments and new formations

The transition from the old to the new working class is not an abrupt one, and the trajectory of the inner cities highlights some of the processes of change. Residual groups of the old (and white) working class were left behind in the rush of affluence. Often the elderly, with patterns of life dominated by the lessons and habits of the past, regulated by the routines of shop, pub, club or chapel, preserved – with little choice – some of the old ways, even as the conditions which had supported that culture were eroded around them. As Coates and Silburn (1970) put it in their study of St Anne's in Nottingham, they were the 'forgotten Englishmen' – although they were also the 'forgotten Englishwomen'. Working class culture always had its defensive features – the mutual supports of 'us' againt 'them' – but under these conditions, these defences became both increasingly desperate and nostalgic. The mobility and hedonism of affluence sat uncomfortably with the locatedness and respectability which were powerful themes of inter-war working class culture. In, but not of, this world of change, the old was defended because the new had no meaning or reality. These defensive traditions came to coexist uneasily with more assertive cultural forms which developed in the inner cities. Some areas experienced 'gentrification' – the rehabilitation of old terraces by the new middle classes: mobile, in search of housing but scornful of the new suburbia and its visage of bland respectability. Habitat lived next door to the Coop. This mix of youth and age was one form of recomposition of the inner cities but race was its other – more visible – axis.

Just as the inter-war working class had found the location and density of the inner city suitable conditions for a defensive and supportive culture, so too ethnic cultures, confronted by the increasingly overt racism of white British society, found a suitable base there, and began to develop their own cultural institutions. This intersection of 'new' ethnic cultures with the old geography of class is not coincidental. In the 1950s, migrant labour was wanted for those 'old' sectors of the British economy (textiles, clothing, heavy manufacturing) where the demand for cheap labour to feed the fires of Britain's industrial expansion was at its highest. Necessarily, then, their economic character as cheap labour meant their distribution followed the pattern of the old industries and their urban centres: Manchester, Leeds, Bradford, the East and West Midlands, and London.

Similarly, within these cities, the residential and social forms for ethnic groups became the old inner-city areas. Excluded from council housing and from the 'good' areas of private housing through a variety of racist mechanisms, they found that some of the old housing of the inner cities (like jobs in the old industries) was available because it was being left behind by the mobile and affluent sections of the white working class. In the Handsworths, Toxteths and Brixtons, the decline of 'traditional' working class culture took place alongside the growth of ethnic cultures. The decaying Methodist chapel was overlooked by the new Sikh temple; local cinemas not rescued by Mecca or Ladbrokes to be bingo halls were reopened to show Asian films; the groups of youths hanging about on street corners became dreadlocked West Indians rather than white skinheads. In these ways, then, the diversity of ethnic cultures came to be overlaid on the geographical pattern of old working class cultures.[11]

Meanwhile, what of the newer, more mobile working class – off to the New Towns and new estates? It would be misleading to see this movement as a simple and total surrender to the new privatized consumer society. No doubt there was an increase of 'home-based leisure' and a rise in the purchase of consumer goods. But the conditions of class are not so easily removed. Affluence, while it lasted, was always fragile – hedged by the possibilities of sickness and unemployment, and underpinned as much by the increase in hire purchase as by rising incomes. The 'never-never' was the route to the consumer paradise. Here, too, traces of the old ways persisted – drink, gambling and sport remained the predominant male working class leisure pursuits. And new forms of cultural difference were constructed out of the new conditions: working men's clubs 'modernized' themselves in the image of market taste in decor and entertainment while stubbonly retaining their distinctive membership structures, internal democracy and institutionalized male dominance.

In surveys of the new leisure society, the working class consistently participate less – retaining a degree of cultural separation even in television viewing. BBC-2 still encounters a stubborn resistance to Reithian cultural improvement, only winning working class audiences for its forays into televised sport. The patterns of life are being remade, but hardly completely anew. Aspects of old structures persist along with traces of old cultures. Women's changing position in paid employment has not undermined their primary responsibility for domestic labour – the re-creation of labour power.

Equally, the increasing mechanization of domestic labour has not abolished its gender specificity. The home remains the site of the woman's 'labour of love'.

It is too early to tell what 'new' cultures may emerge around this new geography. The old patterns have been undermined, but new ways of life have not yet solidified into distinctive cultural shapes. This is not particularly surprising. The old cultures stood in the way of a very powerful vision of Britain's future – an expanding economy, full employment, high wages and the consumer paradise. But that dream has turned into a nightmare. The economic, geographical and cultural consequences of boom do not fit easily with those accompanying recession. One example may suffice to show the speed of this turn around.

In the 1960s, the search for the 'new' working class focused on the car workers. They – the affluent, mobile, mass-production workers – were the vanguard of change and progress. Fifteen years after *The Affluent Worker* was published, this identification of car workers as the advance guard seems true only in the most cruelly ironic way. The car industry (and its workers) has faithfully reflected the dynamics of British capitalism: closures, redundancies, increasing proportions of production transferred overseas, the substitution of new technology for labour, and British plants becoming assembly rather than manufacturing enterprises. Halewood, Dagenham, Longbridge, Ryton, Linwood, Luton and the new estates surrounding them may still be the 'vanguard' – a signpost to the future. But the future which they now represent is a long way from the self-confident assertion of affluence and the end of class. It is a future of decline.

In search of the new Britain

This chapter began by tracing the way in which class and class cultures were embedded in the geography of British capitalism. Nation, region and locality had contours along which stable and well understood class boundaries took shape in the inter-war period. I have tried to show how changes in post-war British capitalism have transformed these once stable patterns. The economic composition of class has been changed; the patterns of class cultures have been disrupted; and the geography of class and culture has been restructured. Nor have these processes of economic, cultural and spatial rearrangement come to a standstill. No new certainties have emerged to replace the old familiar patterns.

Nostalgia for a sense of place that once was is built into this dynamic of change, as the disruption of the familiar is softened by a sense of the past. The new geography of difference may acquire its own stable cultural patterns and boundaries, but as yet, the contemporary cultures of difference are an unstable and uncertain mix of the old and the new.

In this process of change, there is no simple replacement of old cultures by new cultural forms. As was indicated earlier, old patterns are defended and maintained in the face of destabilizing social experience. Resistance to the threat of the new takes many forms, ranging from trade union defence of jobs and working practices in the face of new technology to the maintenance of church and chapel attendance in the face of growing secularization. At other points, old cultural patterns cross-cut and slide into emerging ones – the growing involvement of women trade unionists in previously male-dominated organizations, or the growth of new religious activity (the temples and West Indian fundamentalist Christianity) alongside the established practices of church and chapel. This process is one in which resistance and negotiation are combined in the transformation of cultural forms and practices.

In all of these processes, place – the social geography of Britain – has played a central role. Economic change has had a distinctive geographical shape. As before, Britain is divided along the North–South axis but it is not the same pattern of difference. Scotland, Wales, the Northeast, the Northwest, the Midlands are marked by the process of decline, while the new hi-tech industries in the Southeast and East promise a new economic miracle. The old urban centres have borne the brunt of the dynamics of both expansion and contraction – losing jobs and population to the peripheries and leaving the inner city as the desolate symbol of economic and social decline.

This changing geography has not simply reflected economic change, it is itself part of the process of change as capital and labour face one another in new locations, each bringing their own distinctive history to bear upon the direction of cultural change. The geographical attachments of class cultures have been fractured, and new cultures of class may emerge, attempting to solidify a new sense of place on the shifting sands of British society in the 1980s.

Notes

1 This article draws heavily on collaborative work with Chas Critcher, and has benefited from discussions with John Allen and Doreen Massey. I am grateful to all three.

2 Bennett (1983) and Thompson (1983) provide suggestive analyses of the cultural 'pleasures' of Blackpool. As far as I know, Eastbourne has not received comparable attention.

3 The 'mass culture' thesis can be found in Rosenberg and White, eds. (1957). Hall and Whannel (1964) provides an early critical response.

4 The studies cited here are: Young and Willmott (1962); Dennis *et al.* (1969) and Hoggart (1959).

5 For analyses of the relationship between youth and class see Hall and Jefferson, eds. (1976).

6 For a fuller discussion of this point, see Critcher (1979) and Clarke (1979).

7 Massey (1983) traces this reorganization of labour and its geographical consequences.

8 Damer (1983) analyses the economic and social policies affecting Linwood and their consequences for the local working class.
9 Hamnett (1983) provides a survey of changing housing patterns in post-war Britain.
10 A fuller analysis of the changes in consumption and the market is provided in Clarke and Critcher (1985), chapter 4.
11 Hall *et al.* (1978), chapter 10, and Rex and Tomlinson (1979) provide fuller accounts of ethnic cultures and their place in the inner city.

References

T. Bennett (1983). "A thousand and one troubles: Blackpool Pleasure Beach", in *Formations of Pleasure*. Routledge and Kegan Paul, London.

J. Clarke (1979). "Culture and Capital: the post-war working class revisited", in Clarke *et al.*, 1979.

J. Clarke *et al.* (1979). *Working Class Culture: studies in history and theory*. Hutchinson, London.

J. Clarke and C. Critcher (1985). *The Devil Makes Work: Leisure in Capitalist Britain*. Macmillan, London.

K. Coates and A. Silburn (1970). *Poverty: The Forgotten Englishmen*. Penguin, Harmondsworth.

C. Critcher (1979). "Sociology, cultural studies and the working class", in Clarke *et al.*, 1979.

S. Damer (1983). "Life after Linwood?", paper to British Sociological Association Annual conference, April 1983.

N. Dennis *et al.* (1969). *Coal Is Our Life*. Tavistock, London.

J. Goldthorpe *et al.* (1969). *The Affluent Worker: 3 vols*. Cambridge University Press, Cambridge.

S. Hall and P. Whannel (1964). *The Popular Arts*. Chatto and Windus, London.

S. Hall and T. Jefferson, eds (1976). *Resistance through Rituals*. Hutchinson, London.

S. Hall *et al.* (1978). *Policing the Crisis*. Macmillan, London.

C. Hamnett (1983). "The New Geography of Britain's Housing", *New Society*, 15 December.

R. Hoggart (1959). *The Uses of Literacy*. Penguin, Harmondsworth.

P. Johnson (1964). "The menace of Beatlism", *New Statesman*, 28 February.

D. Massey (1983). "The shape of things to come", *Marxism Today*, March.

J. Rex and S. Tomlinson (1979). *Colonial Immigrants in a British City*. Routledge and Kegan Paul, London.

B. Rosenberg and D. White, eds. (1957). *Mass Culture*. Glencoe Free Press, New York.

G. Thompson (1983). "Carnival and the calculable", in *Formations of Pleasure*. Routledge and Kegan Paul, London.

M. Young and P. Willmott (1962). *Family and Kinship in East London*. Penguin, Harmondsworth.

4

The spaced out urban economy

MICHAEL BALL

Introduction

It is commonplace to anyone that has ever lived in or seen a large city that you have to travel to do almost anything. Going to work, to the shops, to the cinema, a restaurant, to the hospital, school, community centre – then back home – are just a few of the daily activities that involve some form of travel. The journeys themselves could be made in different ways: by walking, by cycle or car; or by some form of public transport, like the bus or train. And as you travel around a city, it is equally obvious that, to varying degrees, activities are functionally clustered. The city centre has offices and the principal shops; there are factory districts; warehousing centres; residential areas separated by social status and date of construction; an entertainments district; out-of-town megastores, and so on. Once you look in detail at these broad, functional spatial separations, even closer specialisations emerge. The City of London, for example, is often just regarded as the financial centre, but within it there is a myriad of specialist centres: legal, printing, jewellery, insurance, banking, stock dealing, shipping and commodities dealing. Once such spatial specializations emerge, they have a remarkable tendency to survive for a long time. Yet they are not immutable, as the waves of deindustrialization that have hit Britain have shown.

Empirically it is difficult to deny that space matters when looking at economic life. Even the most casual observer of a city has to conclude that spatial differentiation is important for the way in which economic activities in a city operate. Yet translating that empirical obviousness into a coherent economic analysis of cities is not so easy.

Two types of question can be asked about how economic processes work across space. The first is: what does the barrier of distance do to the economic activities that have to confront it? This type of question leads to the general issues of why cities exist and how activities within any urban area relate together. The second type of question recognizes that such abstract notions of space are not enough. It asks the question: what does the existence of an already entrenched spatial differentiation of economic and social life do to the current operation of economic activity?

Space is not simply a friction which economic activities have to overcome.

Instead, spatial differentiation also implants a fixity on economic activities. Cities are where they are, and take the historic forms they have because they cannot be moved round at will. This is what makes analysis of the effects of spatial differentiation on the economic life of cities so difficult, yet so fascinating.

I shall start with the features that draw the network of economic activities in an urban area together. Once having looked at them, the chapter moves on to some of the equally important features of space that make the economic life of each urban area so distinct.

Why towns?

A good way to start considering the spatial relations associated with urban life is to ask a basic question: what is the economic rationale of towns? Why do all the economic activities associated with urban life cluster together, rather than spread out across the countryside? The answer obviously has something to do with the barriers created by space. Two types of enquiry can be made into the advantages of such clustering. One concerns the clustering together in reasonably close spatial proximity of the different facets of urban life: workplaces, homes, shops, and transportation networks. These different facets of urban life are *functionally interlinked*.

A description of the emergence of a township around a new steelworks, for example, brings out the interlinkages. First, housing for the steelworkers is built along with shops and rudimentary entertainment and community facilities. Businesses set up to sell to and service the population and the steelworks itself. Other firms might also be attracted by the labour force pool that emerges in the town and by the developing transportation links with other parts of the economy. Eventually, as the town grows, employment in the steelworks might become only a relatively small part of the jobs available in the town.

Such stories can be told particularly for nineteenth-century and early twentieth-century towns, when low incomes and limited transportation systems made travel difficult for much of the population. The emergence of large governmental or military centres produces similar results (Washington DC and Canberra, Australia, being two good examples). Docks and transportation nodes would add other case studies. But each of these examples does not explain why so many similar activities cluster closely together in urban areas. They ignore, in other words, one of the major pulls of urban areas: *agglomeration economies*, which are the second general explanation for the existence of towns. Agglomeration economies refer to the economic benefits activities gain from clustering together. Agglomeration economies bring out clearly some of the causes and effects of spatial differentiation in urban areas, so they will be considered prior to a more detailed examination of functional interlinkages.

The pull of agglomeration economies

Spatial differentiation creates a communications barrier. If two people are working at different locations, and need to talk face-to-face, one of them (or both) has to travel to their meeting place. If they have to be in frequent direct contact, there are considerable advantages in being permanently located near each other, as it avoids the costs and inconvenience of many long journeys. It similarly costs time and money to ship goods around. So suppliers of parts to a factory might find it expedient to locate near to that factory. So far only two examples of agglomeration economies have been specified. In fact, such benefits are extremely varied, and depend on the particular activity in question. Many, though not all, arise from the nature of exchange across space. The trading of commodities, in other words, reacts to the information, time and money costs of distance by creating clusters of economic activity.

Agglomeration economies start to reveal some of the economic mechanisms that structure activities in a city, and show how important spatial differentiation is to them. The benefits of agglomeration can be understood only in the context of their opposite, namely the potential gains to be derived from using distance to minimize the threat of competitors or to avoid the high land costs of locations where many activities want to cluster. So discussion of agglomeration economies will have to look at the push and pull of locational attraction. Moreover, the success of agglomeration economies can lead to their own undoing by creating subsequent diseconomies in the form of high rents and congestion.

Retailing illustrates the attractions and disattractions of clustering together. Customers have to travel to shops, so the further they have to go the stronger must be the pull of the shop, either because it is cheaper or because it offers services not available at closer locations. The costs of crossing space create two pulls on the location of shops. The first is that of monopoly. Benefits can be derived by having a relatively captive clientele that has to travel long distances to buy at a competitor's outlet. The traditional corner shop is a mini-example of the monopolistic position retailers can gain because of the barrier of distance. It takes time to travel to the next shop, so shopkeepers can pitch their prices accordingly, and survive on low turnovers which do not gain the economies associated with the high volumes of the major multiples. Yet such monopolistic advantages frequently are outweighed by the convenience shoppers find in having shops clustered together. In the local shopping street, the benefits to the retailer of clustering can be combined with those of monopoly when the shops sell different things.

Other retailers are far more concerned to maximize the catchment area from which they draw their customers than to use the barrier of space to create a localised monopolistic advantage. Out-of-town stores take advantage of low land prices and good road networks to attract customers from a wide area. Their low overhead costs combined with the economies derived from a high turnover enable them to lower prices below those of their localised smaller competitors.

City centres, as the 'central place' of minimum average travelling time from all points in an urban area (and beyond), are ideal locations for specialist shops and department stores. Both need a large catchment area to generate sufficient custom. The city centre and suburban shopping nodes, therefore, are ideal locations for them. The city centre has particular benefits for the retailers of commodities which customers like to compare before they buy, as occurs with clothes, furniture and hi-fi equipment, for example. In large cities, the central shopping area may develop distinct specialist shopping districts as a result.

Specialist shops clustering together provide one example of the impact of information when it is spread across a number of points in space. Mistakes in the acts of buying and selling can result if an exchange transaction is not made with the best information available. This point can be generalized to many other city centre activities. This is clearly the case for financial markets and commodity dealing. The informal networks which surround such markets become readily available to the individual participant who locates at the centre-of-things. Trends on the Stock Exchange and world money markets need to be known fairly instantaneously. Most activities that have a high element of fashion consciousness in their products similarly benefit from clustering in the city centre. Clothes designers, the media and the entertainments industries find it difficult not to have at least part of their activities located centrally.

Fixing locational pressures

The agglomeration economies discussed so far have placed emphasis on economic advantages to individual firms. All such arguments can suggest is that firms gain advantages from clustering together. Yet the clustering can take place at any location. Nothing that has been said so far *fixes* the agglomerations at particular points. The historical evolution of an urban area and the permanence and fixity of its built environment turns such location-free pressures to agglomerate into actual pressures to locate at a specific point in a particular city.

A cumulative inertia builds up because, once a cluster of activities has emerged to take advantage of agglomeration economies, those economies can be achieved only by locating there. Apart from the need to locate where others do, the provision of buildings in the district to house those activities and a street pattern and transportation network to service its needs gradually emerge. Their existence continues to create advantages long after the initial impetus for locating in a district has passed.

Transportation networks are one of the most important features creating attractive nodes for agglomeration. Once transportation systems evolve, they tend to be augmented gradually rather than radically transformed. Only with major shifts in the dominant form of transport (say, from the horse-drawn carriage to the motor car), or when governments decide to overhaul intra-urban road systems with urban motorways, or to introduce new mass-transit systems, are there dramatic shifts in pre-existing transportation networks. The sheer cost of radical change in transportation systems implants an inertia to points of agglomeration. Moreover, as central areas generate the greatest congestion, new transport networks are added to feed them, so the inertia is encouraged by mutually reinforcing processes. However, even new investments might be insufficient to overcome an increasing congestion of nodal locations, which can act as a major disincentive to further agglomeration. Such transportation networks can then make decentralization feasible as they enable links to the city centre to be sustained.

The history of technical evolutions in transportation has had an important influence on the structure of cities and the spatial distribution of their economic activities. The slow speed and high cost of local horse transportation in the nineteenth century, for example, forced wholesaling and warehousing activities to cluster centrally near to railheads, canals and docks. Late twentieth-century motorway networks, on the other hand, encourage the decentralization of much warehousing activity. Spacious, mechanized, low rent warehouses at the crossing points of two motorways are now frequently far more attractive than their outmoded forebears in the congested streets of city centres. The relative permanence of the built environment, nonetheless, means that the patterns of urban structure laid down under a now non-existent economic rationale still exert a strong influence on current patterns of economic life. The basic pattern of street and rail networks, for example, may have emerged in past centuries. The buildings of the past often still remain as well. Returning to the warehousing example, most city centres that used to have substantial warehousing districts now see the warehouses that remain converted into offices, studios, shops, restaurants and art galleries; everything, in fact, that goes to make up a Covent Garden in London or a Soho in New York.

The physical inertia of city structures is complemented by a similar inertia in the spatial distribution of people. This complementarity is not surprising as living patterns are determined to a large extent by the distribution of the existing built environment as well as by job opportunities (everyone, for instance, needs somewhere to live). Patterns of population distribution exert another set of historically determined influences on location decisions. Populations need servicing with their everyday requirements (schools, medical facilities, shops, entertainments, etc.), and they constitute a potential workforce. These characteristics are further drives for firms to set up in urban areas. The characteristics of the populations distributed across space also vary. Some localities might have a tradition of a particular skill, whereas others might have pools of cheap, unskilled labour. Such variations attract particular firms to areas where their workforce requirements are most easily met.

Pressures for locational change

Other features encouraging agglomeration could be added to the list given so far. But at this stage it is important to stress that, although the nature of the built environment and its historical evolution generate considerable inertia to urban location patterns, there is still a dynamic process of change continually going on. The built environment gets refashioned, slowly but surely, by both the state and private development interests. Populations shift through the cumulative process of thousands of different moves. Accumulation processes in industries lead to changes in production methods and in workforce requirements.

The importance of changes in methods of production and the means by which they occur place agglomeration economies in context. The significance of agglomeration economies depends on: the production process in question; the degree to which it can be segmented spatially; and the conditions of commodity exchange in which production is placed – including workforce requirements, as well as the buying of other inputs for production and the selling of the final output. The changing location patterns of warehousing, for example, have already been mentioned. They are a case where the pull of centre city locations has been severely weakened over time, so that the propensity to cluster has been transferred to other non-city centre sites. Cases arise with other activities where changes in production involve a segmentation of its location. In this way, the new production methods may enable the firm to take advantage both of centrality *and* of locating elsewhere.

Certain types of office employment illustrate the potential advantages that can be derived from such a strategy. Most benefits of locating in centre city offices for, say, a large multinational corporation or a financial institution relate only to certain functions. For the administrative headquarters of a large multinational firm, the benefits of personal contacts, international travel connections, the cultural and entertainments milieux and the prestige of the big city address are essentially characteristics affecting a relatively small layer of management and its financial support staff. Yet often such management functions can only be spatially separated from the other routinized, clerical functions of administration by incurring substantial organization inefficiencies and high travel costs. The high rents and the greater labour costs of centre city locations may consequently have to be paid for all office functions, even if only a few of them actually gained benefits from locating there. A similar argument holds for many financial activities.

Locational options are at the centre of the choices open to a firm when changing its office technology. Computers, for instance, replace administrative staff, reducing the overall cost of an administrative operation staying in a city centre. Alternatively, some activities can remain at a much reduced city centre headquarters, whilst others are decentralized to take advantage of lower rents and cheaper labour costs. The enhanced ability to separate functions with the advent of new computer-based office technology could lead to a firm setting up a number of offices at different locations to tap specific advantages: say, a pool of cheap, part-time female labour for routine administrative functions in one office, and a research and development centre located elsewhere in an attractive 'rural' environment to attract scarce research staff. Alternatively, the whole operation could be decentralized. Each option has advantages and disadvantages which vary for the precise activity in question. It is fairly obvious, however, that the choice of option influences the content of the technology adopted. The new spatial patterns that emerge are not, therefore, simply results of the technology in question but influence its form and the potential future developments it can take.

Functional linkages in the urban economy

Clearly patterns of economic life in cities do not happen by chance. One important systematic component is the functional relations between different types of economic activity. When *describing* such interconnected activities, it does not matter which component you start with, as the linkages are not linear but constitute complex, multi-connected, never-ending circles of production, exchange and consumption. The factories of industrial capital, for instance, need a transport system to ship their inputs and outputs around, and a workforce to produce their commodities; those workers, in turn, need housing, places to buy food and the other necessities of life; the need for schools, hospitals, libraries, public transport and assorted welfare services encourages state intervention; private agencies emerge to service the population; factories develop to provide commodities for those private agencies, for the state, or directly to consumers. Such interconnections, of course, extend beyond the boundary of a single urban area to the world economy as a whole. Some, obviously, have to be localized and limited to the city and its environs: those elements associated with the built environment, such as housing, hospitals, and transport systems are obvious, but not the only case of such necessary localization. In addition, the larger the town the greater are the number of functions that take place within it and, hence, the greater the interlinkages at the urban level itself.

Descriptions of functional linkages at the urban level can be extensive and complicated, but there are likely to be descriptions with which more or less everybody agrees. Theories, however, are needed to explain how and why those particular linkages arise. There has been much disagreement over the explanations that have emerged (and the theories underlying them). Many analysts have interpreted the functional linkages that exist at the urban level as separable from the wider economy. There is consequently said to be an identifiable urban economy and theories of how it 'works'. Alternatively, it has been suggested that the urban areas have one prime function, the maintenance of the workforce, and that this general economic function can be understood only by looking at urban areas.

No theory which gives primacy to a particular spatial level seems very satisfactory. The closure of a steelworks, for example, might decimate the economic life of the town in which it is situated, but the cause of the closure is as likely to be a world slump in steel demand as much as any locational factor. Yet locational factors in the broadest sense must be part of the explanation of why it was that steel plant which closed rather than another. The reasons obviously depend on the precise historical circumstances. The works might have been full of outmoded plant, for example; or the workforce may have been particularly successful in resisting management speed-ups and changes in work practices in the past; or the plant may be part of a company that is in financial difficulty as a whole, whereas it would have survived as part of a financially stronger enterprise. But to explain the closure of that steelworks, spatial differentiation has to play a central part in combination with the general level of economic activity. It is not possible, in other words, to give primacy to one spatial level. If this is so for one urban economic activity, it is also true for others and, hence, for the linkages between them. Assertion of the primacy of particular urban linkages, however, has artificially to close off such complex reactions between different spatial levels.

Even issues associated with state provision have such similarly complex spatial interlinkages. The spate of closures of much needed hospitals in Britain in the 1980s has been undertaken by local area health authorities with substantial implications for the spatial distribution of health care facilities. Yet explanations of why those hospitals went has to range over a whole variety of issues, some local and some not. The physical state and size of the hospital, the population it served, the degree of local mass political mobilization, the politics of the health authority itself are all localized elements of the explanation. Yet interspliced with them has to be an understanding of why the health cuts occurred and the reasons for the forms they took. In part, such questions concern administrative structures and powers, such as the legal powers Ministers of Health have to direct health authorities, but other issues are raised as well. Why, for example, were cuts directed at service provision, instead of at the procurement policies of the National Health Service (NHS), and at the profits of the drug companies from whom the NHS is forced to buy? (Many services and hospitals had disappeared before any attempt was made to reduce these costs.)

The point of both of these examples is to suggest that it is very difficult to understand any economic event occurring in an urban area by limiting the explanation to one spatial level, such as the 'urban'. In this sense, there is nothing that can be conceptualized as the urban economy (in the jargon, it is not a theoretical object). The term is perhaps a useful descriptive one but its role cannot be extended to provide the basis for a theory of the urban economy. In one way or another, to do so is to grant an unfounded primacy to one spatial level.

But, just as it is impossible to limit economic analysis to the urban level alone, it is also impossible to substitute a simple spatial hierarchy instead. Explanation cannot cascade down a spatial hierarchy, starting with developments in world capitalism and ending up with some particular urban locality. Such approaches treat spatial differentiation as essentially residual, as something to be considered once the big stuff is out of the way. This ignores the complexity of spatial linkages, suggested earlier. In particular, it ignores by default the historical and spatial contexts in which global events, like the accumulation of capital, are played out.

To point out the weaknesses of urban theories which start off from the idea of an interlinked urban economy does not mean, of course, that more adequate theories should ignore those linkages. Two points, in particular, can be made at this stage: one relates to the fact that linkages do not necessarily emerge, and the other to the role of space in understanding the linkages that exist.

(i) *The absence of functional necessity*

It is easy to describe, as was done earlier, the evolution of a town in terms of a large industrial plant setting up. Such a pattern of urban growth, however, describes a likely pattern of development rather than an inevitable one. In it, the functional linkages did emerge but there was no explanation of *how* they developed. Once that question is asked, it can be seen that often the functions do not emerge, or do so only in a limited way. There are numerous cases when, say, a housing estate is built with few or no shopping or community facilities provided, or with an inadequate bus service. Similarly, there might be insufficient housing available for all the workers needed by the factories in a town: a problem which plagued many industrial centres in Britain in the 1950s and 60s.

Although the notion of function linkages has been couched in somewhat neutral, almost technical, terms, absences in those linkages in fact create differential social effects for each social class (and groupings within classes). A housing shortage in a locality, for instance, might be a problem for capitalist employers, as they cannot get the workforce they need from the available population. But the shortage obviously affects that population itself much more directly in rising housing costs, long waiting lists and poor housing conditions. Moreover, it is those in the weakest position in the housing system that lose out the most.

(ii) *Functional interlinkages across space*

The linkages between the different aspects of urban life imply that the spatial dimension is crucial to them. Yet, while some activities must interlink within a limited urban area, they have to compete with each other for space within that city. No two activities can share the same locality, each has to be at a different point in space. Spatial differentiation within an urban area is a physical necessity. That physical necessity gives the owners of land and property within the urban area enormous power as they own the sites and buildings where any activity has to locate. So the competition for land leads once again into the question of *how* activities that exist in urban areas get their bit of the city provided for them.

Spatial differentiation within urban areas creates the possibility of an unequal distribution of facilities across space. Some areas might have good health care facilities and schools whilst others are poorly provided. Such differentiations are important down to quite local districts, given such boundaries as the demarcation of school catchment areas. As social groups are similarly arrayed unequally across space, localities of multiple deprivation are likely to arise. The less spatially mobile are likely to suffer the most from such an uneven distribution. Again, they are likely to be lower income families. Households without cars cannot afford, or public transport networks make it physically impossible, to travel to the cheapest shopping districts, or to range across the urban area looking for housing or for work. It makes little difference to an unemployed, low wage manual worker living in an inner city district that equivalent, low paid work is available in some suburb. Housing costs are likely to make it impossible to move to the suburb and transportation costs make it difficult to travel there from his/her present home.

The picture of functional interlinkage within an urban area, therefore, is complex. Specific social groups mesh in with particular activities, yet that meshing requires linkages to be made between specific districts of the city. Problems arise which can affect the functioning of the whole city. These problems are historically contingent. An example from modern office activity illustrates the point. The office uses that push up ground rents in central city areas rely on the existence of a large, low wage, manual workforce to service the buildings themselves, and to attend to the daily needs of office employees in the transportation systems, shops, restaurants, bars and cinemas used by them, and literally to clean up the city centre ready for another working day. Yet increasingly the demands on land by these offices and their effects on land costs make it difficult to sustain even the meanest of centre city accommodation for that low income workforce. Long journeys to work result or labour shortages arise despite high unemployment elsewhere.

Functional linkages across urban space may cut across different political jurisdictions. An urban agglomeration may have a number of local government areas within it, and a hierarchy of metropolitan government, as has existed throughout Britain since 1974. Other public service agencies, such as those associated with health, the police, transportation and water, may also have different geographical boundaries. The ability of local government and other public agencies to provide services and intervene into particular urban problems is constrained by the existence of such a political geography.

In some instances, distinct local municipalities become arenas of social conflict with certain social groups using their control of them to exclude others from the locality or to keep their tax bills low at the expense of others. The large number of suburban municipalities surrounding most large cities in the United States are used in this way, for example. Devices such as zoning ordinances ensure that lower income households cannot move into a wealthy area, while suburbanites gain the benefits of using city centre public facilities without having to pay for them or for the other services provided by the central city government. Successive local government reforms in Britain have limited such a fragmentation of local government (although the proposed abolition of the metropolitan authorities threatens to enhance it again). Yet still there are clear examples of such political manipulation at the local level. Suburban local authorities under pressure from their local populations have managed successfully to forestall widespread development of suburban council housing since 1945, while London provides one of the best known examples in the field of public transport, when an outer London borough, Bromley, forced the Greater London Council to reverse its cheap fares policy in 1981.

The social creation of the built environment

Consideration of how the activities of an urban area get fixed in space requires an examination of the way in which buildings and the physical infrastructure (i.e. the built environment) are created. Many urban theories see urban areas simply as places where things happen, and the built environment as just a passive backdrop. But the built forms of cities have to be produced and maintained as much as the activities that take place within them. Under capitalism, moreover, enormous revenues are appropriated through the provision and ownership of buildings: by landowners, property developers, landlords and other building owners, financial capital and the state (especially via property taxes).

The built environment gets created and reproduced in many ways. If we look at the agencies instigating new building work this variety can be seen clearly: central government builds the main roads; other public authorities (with central government approval) build hospitals and water, sewage, gas and electricity transmission systems; whilst local authorities build schools, community facilities and council housing. The public sector generally builds the infrastructure and public facilities around which private developers build offices, shops and most housing. As private developers need the state to provide the necessary infrastructure for their construction projects, the state becomes involved in one way or another with most facets of urban development. In modern Britain, this interface between the state and the private developer is often mediated through the land-use planning system.

Looking at the instigators of development alone is not sufficient to tell us how development takes place, because a number of social agents have to be brought together to get building done. Land has to be acquired, finance obtained, and the project constructed. There are, in other words, a series of social relations of building provision rather than simply different types of building product and owners of them. Housing illustrates the variety of social relations that can arise.

Modern-day owner occupied housing in Britain is mainly built by speculative housebuilders. They acquire land from private landowners and need planning permission from the state before development can start. New owner occupied houses are sold on the general market for owner occupied housing, where they have to compete with existing owner occupiers selling their houses, and with conversions from other tenures. House purchasers, moreover, generally have to obtain mortgages to finance their purchases. So the social relations of owner occupied housing provision involve relations between private landowners, speculative housebuilders, building workers, market exchange professionals (such as estate agents and surveyors), mortgage finance institutions (particularly building societies), and owner occupiers as purchasers and sellers of housing.

Council housing, on the other hand, is associated with very different social relations of provision. The relationship to landowners has some similarities, as local authorities try to assemble sites on the open market. Powers of compulsory acquisition, however, make it difficult for a recalcitrant landowner to hold out for an exorbitant price. To finance their housing projects, councils are required to take out sixty-year loans at interest rates set by a combination of administrative procedures and contemporary market rates. The building of council housing is generally undertaken by capitalist building contractors, who employ other building firms and workers in complex patterns of subcontracting. Sometimes councils employ their own direct labour departments instead of contractors. Tenants enter council housing through administrative procedures (waiting lists, housing points, etc.) and are subject to similarly set rent levels and management procedures.

Both the detailed content and the overall nature of present day owner occupied and council housing provision are historical products. The broad lines of each evolved during the inter-war years, but considerable detailed change has gone on since then.

The historically specific social relations associated with a type of building provision can be called a *structure of building provision*. Each one has important consequences for the spatial organization of a city and whether any particular function, described earlier, is provided or not. Speculative housebuilders, for example, like to build on suburban greenfield sites to avoid the costs of site clearance and servicing in inner city areas. Precisely where and how much they build depends on the power of landowners, the inducements of the planning system, and the contemporary state of the owner occupied housing market. In the inter-war years, the result was suburban sprawl with minimal community facilities. In the 1980s, their developments are spatially more spread out and much smaller in scale. Council housing, on the other hand, has been associated with inner city clearance schemes and large scale suburban overspill. Considerably more planned in nature, the extent of council housebuilding, its location and the built form it takes are partly products of what its structure of provision can cope with (contractors, for example, do not seem to be able to build cheaply or, often, very well). But much depends on the political and administrative processes of central government.

The historical development of a city and the spatial patterns that have emerged within it affect the locations of developments in each housing form. The clearing of slums and the building of new council housing obviously can only take place where slums exist. Alternatively, up-market owner occupied schemes generally have to gravitate towards already existing high income areas. The future development of an urban area cannot be understood, in other words, without knowledge of its past patterns of development.

Perhaps the clearest instance of the significance for building provision of the spatial patterns that emerged in the past occurs in the relationship between the existing housing stock and new housebuilding. In most urban areas, new housebuilding is usually only a small proportion of the total housing stock. Much of the existing housing, furthermore, might have been built in structures of housing provision long superseded (e.g. nineteenth-century private rental provision). Yet what goes on in new housebuilding has a profound influence on what happens to the existing stock, which depends on the structure of provision in question. Where parts of the existing stock are municipalized, they might be demolished as part of a rebuilding scheme, or renovated and subdivided into flats, or left as they are through shortage of funds. When the existing stock is taken over for owner occupation, the process is often called *gentrification*, as homeowners tend to have higher incomes than the tenants of the housing in its previous private rental use. Small builders and do-it-yourself enthusiasts convert parts of the existing housing stock for owner occupation (and often the results are far inferior to those of the public sector). Yet, as in the public sector case, the viability of undertaking conversions depends on the cost, quality and location of new owner occupied housing. As where people live is influenced by the tenure in which they live, their incomes and the available housing choices, the relations between new building and renovation have considerable effects on the class structures of towns.

Spatial differentiation, however, is not just about how the creators of new buildings have to come to terms with geography and the urban patterns and built environments created over time. Spatial factors also influence the power relations within building provision. Spatial differentiation gives private landowners a monopoly over plots of land. In a context like modern Britain, where the state provides considerable infrastructure facilities which substantially enhance the attractiveness of particular land sites, yet hardly taxes the increments in land values that arise (unless the landowner has a poor accountant), the spatial monopoly power of landowners enables them to extract high revenues from urban development. Speculative housebuilders, to name but one type of private developer, are unlikely, however, to acquiesce to the loss of profit that results from landowners appropriating much of the gains from development in land prices. To avoid such situations housebuilders try to break down landowners' monopolistic power by building at many separate locations across regions, or even nationally. If one landowner has to be paid too high a price, the builder can withdraw his/her offer and build elsewhere. A land bank of sites at different locations considerably enhances the negotiating power of housebuilders as they are then not forced to buy land, whatever its current price, to remain in business. Such spatial strategies, furthermore, influence the nature of housebuilding firms. Small, local housebuilders do not have the locational flexibility of large scale volume housebuilders. Attempts to avoid high land prices, therefore, have helped to encourage the centralization of the housebuilding industry over the past twenty years into an industry dominated by a handful of large producers.

Spatial differentiation is also important for the ways in which council housing is provided. Council housing, after all, is provision by a *local* authority. The housing problems of local areas, and the politics and policies of their councils, are highly varied. Yet, within the structure of council housing provision, central government has increasingly tried to impose uniformity over the way in which council housing is provided. The options open to individual councils in dealing with their area's housing problems are consequently extremely limited. Out of the variety across space of housing problems and housing politics, the structure of council housing provision has created a deadening uniformity.

The discussion of housing provision has raised again the theoretical point that it is impossible to separate out patterns of change in an urban area from the wider economic and social processes of which those changes are part. Spatial factors are intertwined with general social and economic trends in a way in which it is impossible to separate one out from the other.

Emphasis has been placed on housing provision when discussing the provision of the built environment because it illustrates most of the main points that need to be raised. Similar points could be made for other types of land-use. One aspect of office development is worth brief consideration as it brings out the importance of spatial differentiation. It is well known that office developments are often financed by pension funds and insurance companies and bought by them on completion. Offices are seen by them as 'safe' investments with a long profile of income returns that match the revenues they need. As a result of those institutions' activity, the rate of return required to make an office development worthwhile (i.e. its expected yield) is low. Moreover, there is evidence that the rate of return varies inversely with the rate of return on other investments, as the institutions switch more funds into property away from less attractive investments. So office booms tend to occur at the onset of downturns in the economy. For example, a record amount of office space was being built in central London during the early 1980s economic slump.

There is a strong locational element in the office investments of pension funds and insurance companies. To minimise risk, they are interested principally in 'prime' developments at central or well-proven suburban locations. The yield on offices in these areas, therefore, is likely to be the lowest. The overall result is that the investment requirements of these financial institutions have an enormous influence on the amount of office development at any point in time and its *location*. The mass crowding of office blocks together in city centres consequently is not simply a result of demand factors, but also from this particular aspect of supply. Paradoxically, other land-uses have to show a much higher rate of return than offices to be built at those 'prime' locations. It is quite likely, for instance, that the public transport systems required by the congestion created by such concentrated office development have to show an implicit rate of return perhaps ten times or more higher than an office block. The competition for scarce urban land is an extremely uneven one!

A theoretical overview

At this stage it is worth drawing together the variety of points that have been made into a broad theoretical position statement.

When looking at the economic activities existing within urban areas and trying to understand their patterns, problems and linkages, a diverse set of factors is being confronted. At the risk of oversimplification, it is possible to group them into three broad types. Economic life in a city is the outcome of three interlinked processes:
— *the spatial division of labour*, i.e. the points where people do waged work, either for capitalist enterprises or the public sector.
— *the spatial distribution of the population*. This refers to the places where people live. Obviously there are some links to the spatial division of labour, as waged workers need somewhere to live within reasonable commuting limits. But this does not mean there is an exact correspondence. There can be localized labour shortages, while not all of the population is involved in waged labour. The unemployed by definition are excluded, so are the elderly, children, full-time students, and women involved in domestic labour alone. The location of these groups may bear only a weak relation to the spatial division of labour.
— *the spatial distribution of the built environment*. The built environment also has a particular distribution across space which may or may not correspond to the requirements of either production or the population at large. One could go through each building type and note the variations in its cost and

availability across space. As structures of building provision are motivated by their own independent drives, it is difficult to expect a movement of building provision towards an equilibrium satisfying the needs of building users. In particular, the longevity of built structures means that their initial reason for existence may have long passed, but that they still stamp their mark on the city in question; sometimes to its benefit, at other times to its detriment. Because of the nature of contemporary structures of building provision, furthermore, it seems more easy to destroy the built environment than to create pleasant new ones. This gives an overwhelming significance to the historical development of the built structures in a city.

A number of points can be made about the three processes and their interlinkages:

(i) The first point to note is the dynamic character of these processes. Each is the outcome of changing and conflicting relations between social agents.

The spatial distribution of the population illustrates this dynamic. To an extent, it is reactive to the spatial distribution of job opportunities and the built environment; in other words, it is the product of the historical development of the society in question. Yet there are less passive trends. The flight from the urban agglomerations, seen both in North America and Northern Europe since the 1950s, is in part a reaction by people to the lifestyles imposed by large urban agglomerations (as has been the partial move back to the bright lights of the city in the 1970s and 1980s). People do have a degree of autonomy over their lifestyles, and the exercise of that autonomy in aggregate affects the spatial distribution of the population. The possibilities of choice, however, are considerably structured by wider social processes. The services offered by the 'welfare' state, whatever their actual nature, have since the Second World War influenced kinship relationships and, hence, where and how people live. Similarly, the widespread introduction of occupational pension schemes for white collar workers plus the growth of owner occupation have created new location patterns of retired people. Spatial mobility, however, is not equally open to all sectors of the population. The least spatially mobile are likely to be those in the lower status occupational groups, the poor, and households such as single parent families.

(ii) There are obviously links between the mechanisms influencing the dynamic of each of the three processes. Yet, when trying to understand the geographical patterns that emerge, it is difficult to suggest that one of the three processes is always the dominant underlying causal factor.

(iii) Changes in any of the three processes take place within a pregiven historical context. There already exists a geography of production, population and built form. The dynamics of spatial change add to and modify that pre-existing geography, rather than start afresh. This historical context is the basis for understanding the place of particular urban areas within the dynamic of the three processes. They are the sites where changes are played out. Yet one urban area cannot be seen in isolation.

Conclusion

This chapter has tried to show why and how spatial differentiation is important when trying to understand the nature of the economic activities existing in urban areas. The economic life of a city cannot be treated as if it were either spaceless or disembodied from a long process of historical change. It is not possible, using the old adage, to treat urban economic life as though it existed on the head of a pin. Economists, however, have frequently tried to use economic models of the urban economy which abstract from some or all of the fundamental characteristics of spatial differentiation. Not surprisingly, these models have been rather unsuccessful. Hopefully this chapter has shown why an economic analysis of urban areas cannot proceed without a healthy respect for their geography and their position in wider patterns of spatial change.

Another point, however, has also been emphasised, namely the impossibility of giving primacy to particular types of spatial interlinkage. When looking at economic life in an urban area, you must stick your head over the 'urban' parapet. This chapter has suggested that, theoretically, the notion of the urban economy is likely to hinder rather than advance our understanding of the economic problems that beset life in the cities. The title of this chapter, in other words, has a serious as well as a humorous side.

5

Jurisdictional conflicts, international law and the international state system*

SOL PICCIOTTO

Introduction

Over the past few years there have been increasing conflicts between the U.S.A. and several of her main political allies over the extraterritorial assertion of U.S. jurisdiction. Complaints, mainly from European countries and in particular Britain, about the application of U.S. 'long-arm' legislation outside the U.S.A. have led to retaliation against the application of U.S. laws and to U.S. justifications and assertions that some European regulation of international business is also extraterritorial in scope. From the point of view of international lawyers, technical questions of public and private international law are involved, concerning the devising of adequate positive rules for the allocation of regulatory jurisdiction. From the political point of view, policy conflicts are involved which raise the question of whether the Western alliance can maintain a common policy towards such questions as the use of trade sanctions against the Soviet bloc in a period of economic depression (for example, Woolcock, 1982). Both of these approaches assume that the international state system is essentially a functional one, and that provided the correct technical solutions can be found, it is merely a matter of resolving and accommodating national policy differences.

A different perspective on the matter is provided by looking at the literature on the internationalisation of capital and the growth of the multinational corporation. Some 10 years ago, Robin Murray pointed to the growing "territorial non-coincidence" between an increasingly interdependent international economic system and the traditional capitalist (or socialist) nation-state. Murray based himself on the then nascent research on multinational corporations, which proved a major growth industry in the 1970s and now even has its very own United Nations specialized agency, the Commission

* Source: International Journal of the Sociology of Law (1983), 11–40.

on Transnational Corporations. Paralleling the famous remark by Kindle-berger, "the national state is just about through as an economic unit" (Kindleberger, 1969, p. 207), Murray posed the question "whether…national capitalist states will continue to be the primary structures within the international economic system, or whether the expanded territorial range of capitalist production will require the parallel expansion of co-ordinated state functions" (Murray, 1971, p. 86). He received a rapid, perhaps over-hasty, reply from Bill Warren, who in addition to making some good points about the limitations of Murray's theoretical and empirical treatment of the internationalization of capital, was rash enough to prophesy that since the contradictions between capital and the state are essentially non-antagonistic, new international regulatory measures would quickly be forthcoming: "the tax authorities are rapidly getting control of the internal transfer price problem and it is clearly not going to be long before the central bankers, international organizations and State policy-making bodies chain down the Euro-dollar monster so that it is no longer available to do the bidding of large firms" (Warren, 1971, p. 88).

Murray's provocative question sparked a considerable amount of subsequent analysis and discussion, both on the nature and the implications of the internationalization of capital, as well as the theory of capital and the state. It was readily apparent that Murray's essentially structural-functionalist view of the relationship between capital and the state was responsible for the starkness of the alternative he posed: either capital would outgrow the nation-state and lay the necessary basis for new co-ordinated interstate or supranational structures, or its growth would be contained within the boundaries of existing or merged nation-states. Nevertheless, much of his work on multinationals has been seminal, and his raising of the question of territoriality was important and has not been followed up adequately.

Subsequent debates have emphasized that there is a contradictory process in which the national state is increasingly involved in intervention to ensure the social and economic processes of expanded reproduction of capital, yet at the same time these processes are increasingly transcending the nation-state. The internationalization not only of the circuits of commodity-capital and money-capital but also productive-capital is creating an increasingly internationally integrated world economy and social structure. However, this does not take place as a smooth process of symmetrical interpenetration of capitals to be followed eventually by a merging of social patterns and political superstructures. The internationalization of production itself entails an internationalized socialization of productive labour as well as international patterns of commodity circulation, both of which are as much social as economic processes, and involve developing patterns of international class formation and conflict (Van der Pijl, 1979). These developments and the different specific forms they take often owe as much to ideological, legal and even military interventions by states and through international state structures as they do to technological factors. For instance, as we will consider in more detail below, the specific way in which U.S. antitrust law was interpreted and enforced after 1940 played a major part in shaping the characteristic form of U.S. multinational capital using wholly-owned foreign subsidiaries. Thus the state has both been a contributor to the process of internationalization, as well as being affected by it. In Europe, where the international state system originated and which has also been the source of many of the impulses for its subsequent transformations, the development of the European Community involves new and unique processes for interstate integration, and yet the national states still play a vital role, inter-state conflicts continue, and the creation of a single unified super-state is not envisaged by even the most dedicated Europeanist. Yet the impulse to European integration has been as much political as economic (Holloway and Picciotto, 1980).

To summarize, the changes in the international system involve a contradictory and conflictual process of internationalization *both* of capital *and* of the state; and the international crisis of capital is also a crisis of the international state system.

State sovereignty and jurisdiction

In this chapter, I will examine one aspect of this, the growing problems of jurisdictional definition and conflict between the main capitalist states, especially the United States and Europe. It is not surprising that jurisdictional conflicts have occurred mainly between the most powerful states, and in relation to economic regulation, for it is through such forms of economic regulation that the main capitalist states have been involved in shaping the patterns of internationalization of capital. These assertions of national state power in relation to the structuring of international business, and the conflicts and attempts at co-ordination that have ensued, have been in many ways more important than the other more prominent and formalized mechanisms of international law. The increasingly dense network of international organizations of diverse kinds and powers as well as the proliferation of treaties and international arrangements and agreements covering thousands of matters major and minor, have grown in the last thirty years both to express and to mask the increasing powerlessness of the dominated peoples of the world. However, the problem of jurisdiction is more clearly symptomatic of the crisis of international law and of the international system.

The principle of territoriality of jurisdiction is the cornerstone of the international system based on the nation-state. The transition from the personal sovereign to an abstract sovereignty of public authorities over a defined territory was a key element in the development of the capitalist international system, since it provided a multifarious framework which permitted and facilitated the global circulation of commodities and capital. The independent and equal sovereign nation-state is therefore a fetishized form of appearance, for the world system is not made up of an aggregation of compartmentalized units, but is rather a single system in which state power is allocated between different territorial entities. This is important, since exclusive jurisdiction is impossible to define, so that in practice there is a network of interlocking and overlapping jurisdictions.

Historically, the development of the notion of the sovereign state with exclusive powers within its own territory and unable to exercise jurisdiction within the territory of others was also strongly influenced by ideas embodying an overriding universality of law.[...]This was not in accordance with absolutist versions of national sovereignty however, and was eventually eclipsed by nineteenth-century positivism[...]This led to the voluntarist interpretation which held that since the rules of international law emanate from the free will of independent states, no restriction by denial of a claim of jurisdiction over acts taking place abroad was possible unless based on explicit agreement or a universally accepted general rule.

Positivist definitions of state sovereignty and jurisdiction are more or less adequate for liberal forms of regulation, since they can rely on the separation of prescription and enforcement that characterizes such liberal forms. Hence, standard approaches to jurisdiction distinguish between the jurisdiction to prescribe and the jurisdiction to enforce. Jurisdiction to prescribe can be fairly broad in scope and involve a degree of overlap, indeed this may be inevitable. It has been accepted that jurisdiction is not limited to the territory but can also include jurisdiction to regulate the activities of nationals outside the territory. Even territorial jurisdiction involves overlap, since it requires an assessment of the place where the acts to be regulated 'take place'. This question is traditionally resolved by asking where the 'constituent elements' of a regulated activity take place. Thus the classic textbook example of a person in state A shooting a person in state B is used to illustrate the point that jurisdiction depends on whether the rule being applied concerns the acts and intention of the person shooting or the nature of the damage caused to the victim. It is therefore enforcement that acts as the limit to jurisdictional claims, since the actual organization of states as public authorities with defined powers over specific territory means that to be effective a claim to regulate must either involve persons or property within that territory, or must be acceptable to another state which can assist with enforcement. The territoriality of enforcement is sometimes expressed by saying that administrative or executive jurisdiction is territorial.

This approach inevitably comes under great pressure both from the increasing international interdependence especially of economic activity, and from the transcending of liberal forms of state regulation to more direct interventionism, in which there is far less separation between prescription and enforcement of rules. Increasing international interdependence not only means that more than one state is frequently likely to be involved with a particular activity; it also makes it easier for a state interest in jurisdiction to be made effective through actual control within the territory of some person or property connected with the activity. Thus a multinational company can be fined through a subsidiary or permanent establishment in a country even in respect of actions which it claims took place elsewhere. Of course, conversely it is possible for a multinational to arrange for particular physical actions, such as the signing of a contract or the meeting of a board of directors, to take place wherever it might be convenient.

It is such considerations that have led to the move to redefine the territoriality of jurisdiction in terms of 'effects' instead of a physical notion of where actions take place. Such a redefinition is necessary to prevent the sheer evasion of regulations, for example, by arranging for transactions to take place elsewhere. On the other hand, the 'effects' doctrine can in principle involve very broad jurisdictional claims.[...]

Academic commentators have shown an increasing awareness of the inadequacy of the traditional approach to territorial jurisdiction. Akehurst (1972–1973) acknowledges that as long as the 'effects' doctrine is limited to 'primary effects', it could even be a better means of keeping jurisdictional claims within reasonable bounds than the 'constituent elements' approach (p. 155). Lowe, however, goes further, and concludes that the principle of territorial allocation of sovereignty is now unworkable, and calls boldly for

a refinement of the concept of sovereignty in international law, so that it can accommodate both notions of the independence of states and of the increasing interdependence of states, without losing its coherence as a legal principle.

(Lowe, 1981, p. 281)

To this, the even bolder American response has been that the concept of sovereignty is the wrong starting point (Lowenfeld, 1981).[...]It is plain that the international law principles determining and rationalizing the allocation of jurisdiction between states reflect some of the strains of internationalization which I outlined in the introduction. In the remainder of this paper I will discuss in greater detail some of the specific forms which these strains have taken.

Conflicts over extraterritorial antitrust law

The Sherman Act

An ideological rallying-cry much used by U.S. policy-makers in the international arena is the strong commitment of U.S. law to the regulation of business to ensure competition for the benefit of the consumer and society as a whole. Certainly, modern competition law was born out of the populist movement against the big trusts that quickly came to dominate U.S. business in the last part of the nineteenth-century (Josephson, 1934). While the increasing strength of working class organizations, especially the rapidly growing unions, led to direct and fierce conflicts with big capital, the broader social opposition to big capital among farmers, consumers and smaller property owners of all kinds, led to the passing of the Sherman Act in 1890. The broad scope of the Act, its use against some of the Trusts, and even some initial court decisions seemed to hold out the hope that it could be used to dismantle the Trusts. But this evaporated with its interpretation by the courts as a pro-competition law based on the 'rule of reason', the development of new devices for company incorporation by lawyers, and the emergence of a corporatist consensus between big business and the state embodied in the setting up of the state regulatory commissions. Thereafter, the Sherman Act was caught in the basic contradiction that bedevils all competition law, since striking down agreements between firms that are held to restrict competition has the effect of laying them open directly to market forces whose tendency is towards the concentration and centralization of capital (i.e. increasingly large units and fewer firms in particular industries or economies) by the elimination of weaker firms and growth or merger of others.

The Sherman Act was broadly stated to apply to all combinations or contracts in restraint of trade or monopolies affecting trade or commerce within the U.S.A. *or with foreign nations.* In the first case that arose in relation to foreign trade, Justice Holmes gave a landmark judgment whose echoes have been in dissonance with virtually every subsequent opinion by U.S. judges and writers. In *American Banana v. United Fruit Co.* (1909) 213 U.S. 347, the case involved a dispute between two American entrepreneurs with interests in banana plantation operations in Central America, and the plaintiff complained that the defendant had used a variety of predatory tactics to drive him out of the business, including bribery of officials of the government of Costa Rica which had seized his railway (used to export the bananas). Justice Holmes found the plaintiff's case based on the "startling proposition" that acts causing damage which took place entirely outside the United States could be governed by an Act of Congress, and he stated firmly that the legality of an act must be determined by the law of the place where it took place. Although Justice Holmes' opinion was expressed in sweeping terms, its actual effects were subsequently limited.[...]

In the first fifty years' of its life the application of the Sherman Act to international trade affecting the United States, potentially broad, was in practice confined mainly to monopolization of foreign sources of raw materials in order to push up import prices into the U.S.A. Before 1940 there were only a dozen cases begun by the U.S. authorities relating to international commerce, and all were in this category. Yet the period 1890–1940 was one in which world trade was dominated by cartels, which often had the connivance of governments. The major large firms in the main industrialized countries attempted to control international trade and investment in many industries through various type of cartel agreement. These typically involved the controlling of levels of production and the allocation of markets between member firms, each keeping control of its home markets and those in dependent economies.

It was not until the late 1930s that the U.S. authorities initiated a policy of attacking this type of international cartel. This anticartel policy formed part of the wartime planning for a new framework for post-war international economic relations. Freedom for big firms to compete by selling and investing in each other's markets was seen as the key to avoiding the nationalistic politico-economic rivalries and state protectionisms of the prewar period. Thus the planning of the postwar international institutions which would guarantee a liberalized system of trade, payments and investment (the International Trade Organisation, established in stunted form as the G.A.T.T., the I.M.F. and the World Bank) had to be preceded by the elimination of restraints on international competition embodied in the cartels. After 1940 there began a veritable flood of cases involving some of the main prewar cartels, in which American firms had been involved with European and other companies, especially in chemicals, electrical equipment, and high-technology industries, such as metal alloys and equipment such as gyroscopes and optical instruments. The gathering of information for many of these cases was facilitated by the fact that under wartime regulations the U.S. assets of enemy firms were taken over by a U.S. custodian. Most of these cases resulted in the cartel arrangements being abrogated by agreement, often embodied in a 'consent decree' approved by the Courts.

The most visible legal results of this new policy were the postwar landmark court judgments, of which the *Alcoa* case is the foremost. This case involved the prewar aluminium cartel set up under the umbrella of a Swiss corporation by French, German, Swiss and British monopolies and with the participation of Arthur V. Davis and the Mellons who dominated the U.S. market through Alcoa (the Aluminium Corporation of America) and who had set up a Canadian corporation, largely it seemed to enable them to participate discreetly in the cartel. Since the court held that the existence of a common group of connecting shareholders was not sufficient to ignore the separate legal identity of Alcoa and the Canadian company, it was faced with a conspiracy made up entirely of non-U.S. firms and whose meetings had all taken place outside the U.S.A., even though, of course, it affected the U.S. economy, as part of the world economy. Justice Learned Hand however hesitated little in holding the Sherman Act to apply, on the grounds that "it is settled law that any state may impose liabilities, even upon persons not within its allegiance, for conduct outside its borders which the state reprehends; and these liabilities other states will ordinarily recognise".

This dictum initiated the so-called 'effects' doctrine of jurisdiction. As applied to antitrust law this policy had a significant effect on postwar developments in foreign investment. The broad effect of this policy, which was backed up by strong court judgments, was to invalidate the involvement of any U.S. firm in any arrangement or agreement having the effect of limiting competition, which included not only the classic international cartel but even a joint venture to set up a company abroad.

This policy of application of the U.S. laws had a significant effect on the form taken by the postwar expansion of large American firms in world markets. The sweeping away of the cartels removed any legal barrier to U.S. firms wishing to make incursions into foreign markets, and the doubts cast on the legality of any form of joint venture made a significant contribution to the feature that the postwar foreign expansion of U.S. based firms typically took the form of the 100%-owned foreign subsidiary abroad could be said to involve a 'conspiracy' (between the parent company and its legally separate offspring) and could be argued to 'affect U.S. trade' by e.g. reducing exports from the U.S.A. in favour of local production, in practice the U.S. authorities did not take this view and started no action on this basis.

It was during the 1950s that foreign firms and governments began to object to the application of U.S. antitrust law in circumstances which it was alleged involved 'extraterritoriality'. In many cases the invalidation of agreements under U.S. law affected only the private legal rights of the parties, and the requirements of the enforcement of the U.S. assertion of jurisdiction meant that non-U.S. firms could attempt to evade the effects of the legal cases, if they were willing to do no business in the U.S.A. However, the power of U.S. authorities over U.S. firms meant that the continuation of any arrangements with them was impossible. However, no direct legal conflict arose from the U.S. assertion of jurisdiction unless third party rights or foreign state regulation were involved. One such case involved the British chemicals giant, I.C.I., whose involvement in a number of chemicals cartels with firms including the U.S. firm Du Pont, had been attacked in the U.S. courts. One aspect of the cartel with Du Pont involved patent-pooling, and I.C.I. assigned British patents that it had received from Du Pont to an affiliated company which it part-owned, British Nylon Spinners. The New York court had ordered the mutual return of the patents that had been exchanged by I.C.I. and Du Pont, but the decree (drawn up no doubt by acute negotiations between Wall Street firms) had included a proviso that nothing it required should oblige a defendant to act contrary to any laws or orders of a foreign state "or instrumentality thereof". British Nylon Spinners brought an action against I.C.I. to enforce its rights under the assignment of the British patent, and Lord Denning was only too happy to proclaim that "the writ of the United States does not run in this country", and enforce those assigned rights. He also carefully pointed to the savings clauses in the New York court's decree, which effectively prevented I.C.I. from being subjected to contradictory court orders. However, the U.S. courts subsequently further elaborated this "foreign sovereign compulsion" defence, and required proof of a *bona fide* attempt by the foreign defendant to carry out the order.[...]

Retaliation and attempts to accommodate jurisdictional conflicts

The assertion of jurisdiction to regulate international business by various U.S. agencies led to increasing conflicts, in particular in relation to matters such as shipping liner conferences. These led various governments to pass legislation attempting to prevent U.S. authorities from carrying out investigations which it was claimed infringed the jurisdiction of other states. For example, the British Shipping Contracts and Commercial Documents Act of 1694 empowered the government to prohibit those carrying on business in the U.K. from complying with foreign government measures which appeared to "constitute an infringement of the jurisdiction which, under international law, belongs to the U.K.". Attempts were made through bodies such as the O.E.C.D. to coordinate western countries' policies to matters such as shipping liner conferences.

At the same time, the U.S. also attempted to modify the expression of its jurisdictional claims in ways that most clearly recognise the interests of other states in the regulation of international business. To clarify the applicability of the antitrust law to international business and attempt to mollify criticism, the Antitrust Division of the Department of Justice in 1977 produced a detailed *Antitrust Guide for International Operations*. At the same time, two lower court decisions have reviewed in detail the jurisdictional problem and adopted a modified approach. In particular, Judge Choy in the *Timberlane* case enunciated a "jurisdictional rule of reason" which required courts carefully to weigh the foreign relations impact of an assertion of jurisdiction over international business. On the other hand, some writers have argued that the introduction of the "jurisdictional rule of reason" by U.S. courts has failed to appease other states because the balancing of policy interests of national states in regulating a particular activity "will usually reflect an understandable bias in favour of the forum's policy" (Maier, 1982). Maier argues that the weighing of contracts in determining whether the assertion of jurisdiction is "reasonable" should not be in terms of competing local law policies, but should be informed by the common needs of the international system [...]. However, this approach does assume that there is some sort of political or ideological basis for the formulation of generally acceptable principles that express the "common needs of the international system".

In practice, the increasingly acute crisis of capital has led to further competitive conflicts between giant firms and protectionist interventions by states. The most notable occurred when the U.S. firm Westinghouse defaulted on fixed-price contracts to supply uranium to nuclear power stations it had built, and alleged that world uranium prices had been pushed up by a secret cartel of uranium producers. It instituted private treble-damage actions under the Sherman Act against the mining companies, including the British-based RTZ, claiming billions of dollars in damages, and the U.S. Attorney-General began a grand jury investigation. Other governments, some of whom such as Canada and Australia were reported to have backed the cartel, moved to block the U.S. proceedings. The British government, largely as a result of the Westinghouse affair as well as the American action against North Atlantic shipping conferences, moved to replace the 1964 Act with the Protection of Trading Interests Act 1980.[...]

The basic provisions of the Act are designed to apply not merely to an infringement of U.K. jurisdiction, but to enable retaliation, in concert with other states if necessary, against an extraterritorial assertion of jurisdiction which threatened U.K. interests. Thus persons (including companies) carrying on business in the U.K. could be given instructions, or could be permitted to recover the penal element of multiple damage judgments, even if the assertion of jurisdictioin complained of did not infringe that of the U.K. at all, so long as the Minister considered U.K. trading interests were involved.

This involves two separate possibilities. One is that the jurisdiction of a third state is infringed in circumstances which involve damage to British trading interests. The other is that the initial assertion of jurisdiction, although it is 'extraterritorial' does not involve an infringement of a jurisdiction which could be said to be *exclusively* British. In other words, it provides the basis for a retaliation in defence of British trading interests against an assertion by another state of jurisdiction to regulate international activities that might be considered to fall within the *concurrent* jurisdiction of *both* states. Thus the comments by the U.S. government that the British Act itself involves some 'extraterritoriality' (U.S. Government, 1979) slightly misses the point. The 1980 British Act is important because it abandons the attempt to define mutually exclusive spheres of jurisdiction between states under international law. Instead it recognizes that where one state unilaterally regulates an activity that cannot be said to take place entirely within its territory, other states must establish a competing or conflicting regulation if they wish to compel an attempt at international co-ordination.

Conflicts over embargo policies

Over the same period, conflicts have occurred over other areas of 'extra-territorial' assertions of U.S. jurisdiction. These have mainly involved the application of U.S. business regulations to foreign companies owned or controlled by U.S. citizens or U.S. companies – that is to say essentially, foreign subsidiaries of U.S. multinationals. One significant area has been U.S. embargo policies, initially against the Soviet bloc and China, then Cuba, and more recently the financial sanctions against Iran.

The Cold War embargoes

The history of U.S. attempts to orchestrate western economic warfare against communist nation dates back to the period 1947–50 (Adler-Karlsson, 1968).

Initially, the emphasis was on achieving the economic isolation of the Soviet bloc through co-ordinated export controls by the U.S.A. and its allies. The Export Control Act 1949 stated it to be U.S. policy to use trade as a political weapon, by empowering the government to prohibit the export of goods or technical data which might make a significant contribution to the military or economic potential of any nation or combination of nations threatening the security of the United States.[...]

Using a combination of inducements and threats, the U.S. quickly developed a fairly effective economic isolation of the Soviet bloc between 1948 and 1953, which however also had the effect of reinforcing Stalin's aim of consolidating Soviet domination over eastern Europe.

The emphasis of U.S. controls in this period was on exports from the U.S.A., although the Export Control Act also covered the use abroad of U.S.-origin technical data. The strength of U.S. industry and the greater attraction of U.S. aid to countries that had emerged weakened from the Second World War, could be used to ensure international co-ordination. However, there is some evidence that the extent of European co-operation was made to seem greater than in practice it was, due to the problems of definition and enforcement. Adler-Karlsson in his thorough study of this period concludes that it was largely the threat of the loss of the significant quantities of Marshall Aid that ensured compliance by the European countries.

Amongst the main capitalist countries co-ordination was established in the form of an obscure but apparently powerful mechanism, the C.G.-Co.Com. (Consultative Group–Co-ordinating Committee) set up in 1950. Although the U.S.A. put the initial proposals for this mechanism to the O.E.E.C., it thought that this would be too open and public a body for the purpose, and C.G.-Co.Com. apparently has existed merely as a 'committee' independent of any formal international organization. Almost total secrecy has surrounded it – even its innocuous name was considered 'classified' as late as 1953. It is thought to be based on no written agreement or treaty, but simply a 'gentlemen's agreement' that the lists of embargoed items, agreed on the basis of unanimity, will be enforced by each member state. Members have been the N.A.T.O. countries with the exception of Iceland and the addition of Japan. The main issue throughout has been disagreement over the concept of 'strategic' materials. A strong current in U.S. views has been that any items that help to build the Soviet industrial base assist its preparations for war. Hence, an early disagreement with Europe was over British exports of steel. Vociferous anticommunists in Congress specified that "not even a shirt button" should be exported to the enemy. This led to Khrushchev's satirical response that buttons are obviously the most strategic commodity since without them a soldier would have to fight with one hand, the other being required to hold up his trousers (Adler-Karlsson, 1968, ch. 3). It was after 1953 that the embargo lists apparently began to shrink as the European view that they should cover militarily significant items began to predominate. In 1958 there was a further relaxation. From that point on, there was a significant gap between the U.S.A.'s own total ban, and the internationally agreed embargo, which involved the continual re-interpretation of the ambiguous notion of 'strategic' items.

It was at this point that the potentiality for jurisdictional conflict became a real one. By the late 1950s foreign direct investment by U.S. corporations had begun to build up substantially, especially in Europe. In any case where there was a disagreement between the U.S. and European governments as to whether a particular commodity was 'strategic' and should be embargoed, a dilemma could arise for the U.S.A. and for U.S. multinationals whether a U.S. corporation prohibited from selling an item to a communist country could do so from one of its foreign subsidiaries. This was prevented, both by the extension of export control procedure to cover re-exports and by transaction regulations passed under the Trading With the Enemy Act, Section 5b. This was a typical wartime emergency legislative enactment first passed in 1917, giving the President wide powers in times of war or declared national emergency to regulate transactions or expropriate property. It was revised and activated in the Cold War era, when President Truman declared a "national emergency" to be in effect on 16 December 1950, after the outbreak of the Korean War; this was apparently considered to be still in effect up to thirty years thereafter. Finally, the embarrassment of this permanent state of declared national emergency, combined with the need to rationalize U.S. powers of economic retaliation following the Arab oil boycott of 1973–4, led to the passing of a slightly more restricted empowering statute to replace the Trading With the Enemy Act, the International Emergency Economic Powers Act 1977. In the meantime, foreign assets control regulations were applied under the former Act throughout the 1950s and 1960s, first against eastern Europe and the Soviet Union, then the People's Republic of China, and later North Vietnam and Cuba. These regulations applied not only to persons resident or companies doing business in the U.S.A., but also to any enterprises wherever organized or doing business which could be said to be "owned or controlled" by U.S. nationals or companies. Under the wide definition of 'control' adopted, even a minority shareholding in a foreign company could be held to constitute 'control'. Thus a U.S. company could be warned that a transaction being entered into by a foreign affiliate, even if it was jointly owned with foreign investors or companies, constituted a criminal offence by that affiliate under U.S. law. The Department of the Treasury frankly admitted that the enforcement of these regulations inevitably raised foreign policy questions with America's allies. Inevitably, also, they raised legal problems of jurisdictional conflict (Berman and Garson, 1967).

One notable case arose in France in 1964. Fruehauf-France, a two-thirds subsidiary of the U.S. Fruehauf Corporation, contracted with the French company Berliet to supply trailers for trucks, which were to be supplied to China. It seems that initially this destination was not known in the U.S.A., but once it became known, the Fruehauf parent company was ordered not to allow the trailers to be supplied. It is not clear whether the U.S. multinational covertly sought a way to evade this order, or whether there was a genuine disagreement between the U.S. parent and the French subsidiary's management but, for whatever reason, the French minority directors applied to the French courts for support. In the event, the Paris Court of Appeals took the view that the interests and personality of the French company should be safeguarded even against those of its majority owners, and it appointed an administrator over the Company for three months to execute the contract.

One factor revealed by the Fruehauf case and similar incidents was that the subsidiaries of U.S. multinationals abroad were put at a competitive disadvantage in relation to other firms operating in the same country by the gap between the embargo policies of the U.S.A. and other countries. This became more acute during the late 1960s and 1970s with increasing competitiveness of international markets and increasing pressure, especially in western Europe, for *détente* with the Soviet Union and the further relaxation of trade bans. The American export control legislation was amended several times to relax the procedures, and to make it easier for the substance of the controls to be varied according to the variable winds of *détente* policy. This culminated in the Export Administration Act of 1979, amending and re-enacting the 1949 Act. Crucially, however, the 1979 Act no longer applied merely to the export of goods and technology *from* the U.S.A., but explicitly covered the "export of any goods or technology subject to the jurisdiction of the United States or exported by any person subject to the jurisdiction of the United States" [Section 4(a)(1)]. This provided the powers to apply export control to foreign subsidiaries of U.S. multinationals, on both 'national security' and 'foreign policy' grounds, and even where no national emergency had been declared as required by the International Emergency Economic Powers Act. It was clear that, whatever happened with *détente* policy, the U.S.A. had by no means renounced the economic embargo weapon nor its application as widely as U.S. economic power could reach. Indeed, it is clear that, especially after the Arab oil boycott of 1973–4, U.S. policy-makers were preparing defences against such economic threats to the U.S.A. and for the activation of U.S. economic power on a suitable or necessary occasion.

The Iranian freeze and the Eurodollar

It seems that this occasion was the Iranian crisis of 1979–80. Whatever the background of the deterioration of political relations between the U.S.A. and Iran, it seems that part of the motivation for the scope and type of sanctions applied was to test the effectiveness of U.S. control of international economic weapons. In terms of U.S. apologetics, the justification is that economic warfare saves lives. The main problematic aspect of the financial sanctions was their application to Eurodollar deposits made by Iran outside the U.S.A. Since the loans made to Iran far exceeded Iran's deposits in the U.S.A. alone it was imperative that the sanctions should apply to all Iran's foreign holdings in the international banking system. It does not seem that any attempt was made to co-ordinate such a policy with the western allies, and it seems unlikely that such a co-ordination could have been possible. Instead, the U.S. freeze regulations were unilaterally made applicable to dollar deposits in U.S. bank branches abroad as well as in the U.S.A. In addition, the U.S. banks sought the means to put pressure on other international banks to go along with the policy, by creating a sort of 'domino effect'. The U.S. regulations were amended so that U.S. bank branches abroad were not only required to freeze Iranian deposits, but also were permitted to 'set off' against these frozen deposits any payments due from Iranian entities. The set-off was not allowed within the U.S.A., apparently because a number of smaller U.S. banks were involved in consortia with loans to Iran without having any Iranian deposits. Once the freeze was applied by the U.S. banks the Iranian authorities found themselves short of funds to service the extensive loans negotiated under the Shah; they naturally brought legal actions in European courts, notably in London and Paris, to secure the release of their funds there. In the meantime, the refusal of the U.S. banks to accept instructions regarding these deposits meant that Iran was exposed to potential default on her Eurodollar loans, and indeed the U.S. banks that had led the consortia for these loans took steps to trigger the default clauses in the agreements. This action put pressure on the European international banks involved in the loans also to use legal rights of set-off against Iranian deposits, since once default is declared all participating banks are under an obligation to make their best efforts to seek satisfaction of the loan.

These unilateral legal and banking moves were of course accompanied by a major ideological campaign, as well as diplomatic initiatives to rally America's allies. Despite obvious hesitations, what was obtained was at least silent compliance. The actual sanctions applied by European states were minimal since they did not apply to the performance of contracts concluded prior to the hostage seizure (and in Britain a parliamentary move further exempted contracts concluded prior to the legislation itself). But most importantly, no retaliatory action was taken in respect of what had amounted to an assertion of U.S. jurisdiction over the entire international financial system. Carswell's (1982) retrospective evaluation admits that it was the unique and short-lived nature of the emergency that enabled the U.S. freeze to be effective. [...]

The Soviet gas pipeline

That the acquiescence by the European governments in the way the Iran embargo affected Eurocurrency transactions did not imply any general acceptance of the application of U.S. embargo regulations to the international economy was shown by the conflict that developed in 1982 over the gas pipeline embargo. At the end of December 1981 the Reagan administration decided to tighten the foreign policy embargo applied against the Soviet Union as a means of demonstrating U.S. pressure on the Soviet Union over events in Poland, as much for domestic and international propaganda as anything else. Carter's wheat embargo over Afghanistan had been a mistake, costly both in subsidies for, and in political support from, American farmers.[...]

At the same time, there was opposition from significant elements in the new administration to the involvement of western Europe in the Soviet Union's major project of connecting its Urengoi natural gas field by pipeline to the western European grid, to sell gas, for valuable foreign exchange. Since little of this technology had military applications it would amount to a declaration of economic warfare, and the allies of the U.S., far from being ready to give the necessary co-operation to ensure that competing suppliers did not fill U.S. contracts, had already set their face against any interferences with the pipeline. Nevertheless, the President issued regulations in December 1981 which applied controls on the export to the U.S.S.R. of oil and gas transmission and refining technology (exploration and production technology had already been embargoed for some years).

The main contractors for the compressor stations for the pipeline were European firms (a Mannesmann-Creusot Loire consortium and Nuovo Pignone of Italy), as were the suppliers of the turbines for the compressors (AEG-Kanis, John Brown and Nuovo Pignone). However, the turbines were designed by General Electric of the U.S.A. and manufactured in Europe under licence. Also, some of the components for the turbines were contracted for by U.S. firms: notably GE itself, which under the terms of the licence agreement was to supply rotor blades and moving parts, and Dresser, whose French subsidiary was contracted to deliver compressors. In addition, there was a network of smaller suppliers, some of them subsidiaries in Europe of U.S. firms or licensees of U.S.-origin technology patented in Europe.

The application of the December regulations to the pipeline contracts was far from clear. GE quickly announced that the embargo prevented it from delivering the components contracted for. From one point of view, the European firms were unwilling to court American displeasure and run the risk of losing other contracts based on licensed technology: however, both AEG and John Brown were in severe economic difficulties and the work for the contracts was important for them, in the short run. The U.S.S.R. spoke boldly of finding alternative suppliers or diverting some of its own capacity to build the pipeline.

Negotiations between the U.S. and its European allies took place in the first half of 1982. The European governments expressed their objection to the use of the Polish question to revive U.S. opposition to the pipeline project, and stated that while they were willing to take measures such as tightening of credit for Poland and the U.S.S.R. to express displeasure, they were not prepared to "wage economic warfare", especially not by cancelling the pipeline contracts. With additional pressures from Japan and from U.S. firms linked with the contracts such as GE and Caterpillar, it seemed that the anti-pipeline elements would be forced to retreat. Instead, a National Security Council meeting in June 1982 authorized the President to move forward.

The additional regulations, effective from 22 June, clarified that the embargo applied to goods produced abroad by companies or organisations "owned or controlled" by U.S. citizens, residents or corporations. But they also went further. The embargo applied to non-U.S. companies located outside the U.S.A. in respect of goods produced on the basis of technology patented abroad but 'originating' in the U.S.A. A particularly striking feature of these provisions was the way in which private contractual arrangements made abroad were used as the basis for the imposition of U.S. regulations. This is reminiscent of the so-called 'tertiary' aspects of the Arab boycott of Israel, whereby main contractors for projects in Arab countries have been required to ensure that sub-contractors are not black-listed persons. Ironically, this has been objected to by the United States as an infringement of U.S. jurisdiction, and is the target of counter-boycott actions.

The explicit application of the embargo to European firms and U.S. subsidiaries in Europe quickly led to a highly publicized wrangle between the U.S. and its European allies. Bonn, London and Paris were quick to attack the unilateral application of the embargo to European firms, and began to apply diplomatic pressures in combination with counter-controls on the firms concerned to achieve a reversal of the measures. The governments of France, Italy and Germany made statements during July and August objecting to the embargo and instructing their companies to fulfil contracts with the U.S.S.R. They appeared in general to use no legal power, but merely to write to the companies communicating the government's expectations. The case of Dresser France created some difficulty however. A senior Vice-President of the Dallas-based company revealed that it had instructed its French subsidiary in June not to proceed with filling the gas pipeline orders, in compliance with the U.S. regulations; but the French subsidiary had received contrary instructions from the French government, as well as being the target of demonstrations organised by the French trade unions demanding the continuation of the work. An application by Dresser to a U.S. court for an interim order invalidating the U.S. regulations failed; but the French government then invoked powers under 1959 emergency legislation to requisition the company in the national interest to fulfil the contracts. The U.S. authorities immediately retaliated by declaring a "denial of U.S. export privileges" against Dresser France, the usual administrative sanction against such violations of export controls.

In this situation, the companies did seem genuinely caught in a cleft stick. Although noncompliance with the U.S. embargo would certainly enable the continuation of much-needed work in their plants and the avoidance of contractual penalties, both U.S. subsidiaries and U.S. licensees of U.S. technology were dependent on their U.S. links for much of their work. U.S. sanctions could lead to immediate losses. However, GE and other U.S. firms also benefit from such links with European companies and would lose if the Europeans started to use competing technology, as well as relying to some extent themselves on contracts with the European firms. Clearly the working out of who could suffer most would involve a major trade war. Already U.S. foreign policy had greatly suffered from having converted a show of strength against the U.S.S.R. into a major row with its own allies.

Although the European governments each took their own retaliatory measures, there were obvious advantages in co-ordinating these, since the U.S.A. would be less likely to risk enforcing the embargo by trade sanctions against a co-ordinated European response. Diplomatic co-ordination took place directly between governments; but the European common interest, and the effect on European trade, provided a basis for an intervention by the Commission of the European Communities.

The Commission was obliged to handle the arguments relating to jurisdiction very carefully, because the strongly held views of some member states on extra-territoriality could be interpreted to conflict with the views of the Commission itself, particularly those of its Competition Directorate, which accepts the 'effects' doctrine. European exports for the U.S.S.R. pipeline could hardly be said to produce direct and substantial effects on the United States; so even under the effects doctrine the measures were of dubious validity. This was supported by the fact that they had been adopted under the 'foreign policy' and not the 'national security' provisions of the 1979 Act. Indeed, although the legal argument did not bring out this political point, the gas pipeline had been chosen for the embargo precisely because its suspension would produce little harmful effect on the U.S.A. The replacement of the Carter wheat embargo by the Reagan measures against the pipeline resulted precisely from the convenience for the American ruling class to fight its battles, by using its predominance in the world economy, in a way that caused least economic and political difficulties for the U.S. state and its companies, and most for their allies and competitors.

The eventual retreat from the brink of an internecine trade war took place on terms as replete with ambiguities as were the circumstances that nearly precipitated it. Throughout the crisis, U.S. sources had argued that their unilateral measures resulted from the failure of their allies to agree to more effective measures against the Soviet bloc, notably by tighter credit restrictions and a strengthening of Co.Com. President Reagan's announcement of the end of the pipeline sanctions, in November, stated that it had been enabled by the achievement of agreement with U.S. allies that multi-lateral controls on technology exports and credit facilities would be strengthened and no new gas supply contracts would be signed pending the conclusion of a joint energy study. However, France immediately denied that any agreement had been reached, and European official sources generally stated that no new initiatives were envisaged: the Co.Com negotiations would continue, as would discussions in O.E.C.D. on export credit and in the I.E.A. on energy. Certainly, these bodies will provide the forums for continuing negotiation both of the substantive policy issues and of the jurisdiction question.

Conclusions

It is not difficult to discern the political and economic reasons for the specific policy conflicts involved in the conflicts of jurisdiction that have occurred, especially between the U.S.A. and Europe. In addition to the main incidents of which I have given an outline, the jurisdictional issue has arisen also in relation to other matters, such as the application of U.S. securities and stock exchange regulation. Indeed, no sooner had the gas pipeline matter dropped out of the headlines than another conflict replaced it, this time mainly between the U.S. and the U.K., over the liquidation of Laker Airways, and whether the claim by its liquidator that the airline's collapse was due to a conspiracy between other airlines and airframe makers could be litigated in U.S. courts. The regulation of transatlantic air traffic is clearly an area of concurrent jurisdiction, and the British use once again of the Protection of Trading Interests Act was aimed at blocking U.S. intervention and forcing international negotiations.

One of the reasons for the frequent allegations of extraterritoriality against the U.S.A. has been the rapid expansion of its direct foreign investment, mostly in the form of 100% owned subsidiaries, and especially into Europe, coupled with the American predilection for a formalistic regulation of business, compared to the more informal corporatist concertation often practised in Europe. Nevertheless, European authorities have now been obliged to examine more closely their jurisdictional policies in relation to business regulation, especially where multinational companies are concerned. Americans have not been slow to point to European regulations which could have an extraterritorial scope (see e.g. Vagts 1982). Indeed, in the aftermath of the pipeline dispute, the European Commission has felt it necessary to attempt to co-ordinate internally its policy on jurisdiction, so that its objections to other states' claims do not contradict too starkly with its own assertions of jurisdiction. However, the contradictions and differences of viewpoint within the Community are so great that it is unlikely that any such guidelines on E.E.C. jurisdiction will actually be achieved, let alone be publicly issued as a Statement of Policy. Other ideas such as a European antiboycott statute, also face big political and legal objections. Nevertheless, the Commission has had to take some note of U.S. objections to extra-territoriality in some of its proposals, however spurious some of the objections seem. Indeed, in general, European regulations applying to international business have been framed much more cautiously than U.S. ones. However, U.S. lobbyists such as the American Chamber of Commerce in Brussels, have objected to European Community proposals on jurisdictional grounds, notably the Vredeling proposal for a directive to require multi-plant and multinational firms to establish machinery for consultation with worker representatives; and also the seventh Company Law directive requiring consolidated accounts for groups of companies. In both of these cases the question has been whether a parent company outside the E.E.C. could be obliged to furnish information relating to its global activities, in view of its operations in the E.E.C. through a subsidiary or branch. In such cases it is not difficult to see how the obligations can be laid on the subsidiary in such a way as to leave the parent little choice. Another area where there are continuing objections to jurisdiction is competition law. Notably, the current proceeding by the Commission against the computer giant IBM has involved both the company and the U.S. government arguing that the Commission's attempt to make IBM change its marketing practices to make them more 'transparent' and open to competition from plug-compatible competitors goes far beyond the jurisdiction of the E.E.C.

As I have argued, it is very hard to maintain that the issue in such cases is about the delimitation of an area of exclusive jurisdiction. In practice, the principle of territoriality can be just as broadly applied as the effects doctrine. The underlying issue is that the increased internationalisation and integration of the world economy has created large areas of overlapping jurisdiction, in which conflicts are found to occur unless national regulations are either to be ineffective, or are co-ordinated. Combined with this, the very different nature of the relationship between capital and the state makes liberal forms of accommodation of jurisdictional overlaps ineffective. In the past private law jurisdictional overlaps could be accommodated by conflicts of law rules, and public law was penal and confined to the territory. Now there is a wide range of state regulation of business that is not considered strictly penal (although penal sanctions are applied very often) and that must be applied to multinational business if it is to be effective. The nature of this form of regulation also negates the other method of accommodating jurisdictional conflicts that is most often suggested: to allow a broad overlap of jurisdiction to prescribe while limiting enforcement strictly to territory, since in any case state enforcement agencies have no powers beyond their borders without agreement.

As the examples I have outlined show, sanctions such as the "denial of export privileges" can be very effective against international business even if only applied territorially. It is not so much the territoriality of enforcement that sets a limit to jurisdiction as rather the creation of conflicts with other states' policies or laws. Sometimes there may be conflicts only with private legal rights which may be privately adjusted by the parties or by provisions such as *force majeure* clauses; or the conflict with law or policy may be (reluctantly) tolerated by the other state for a period (as with the Iran freeze). But the move towards the creation of counteracting administrative powers such as the British provisions for the "protection of trading interests" shows that concurrent regulation of international business is increasingly difficult to tolerate or accommodate through liberal forms. Therefore, what is being attempted increasingly is the setting up of direct processes of co-ordination between state agencies, which provide arenas within which negotiations between those agencies and the giant multinationals can be played out. Within such internationalized neo-corporatist state structures it is likely to be the multinational corporations who will dominate, even more so than on the national plane. Nevertheless, there are serious contradictions for capital involved in the disjunction between economic and political processes of internationalisation, and the redefinition of the international state system is by no means an automatic and straightforward process.

Acknowledgements

I am grateful to many students, colleagues and friends for stimulating discussions and comments on drafts of this paper, especially to Julio Faundez B. and to members of the European Conference on Critical Legal Studies. I also obtained valuable help from officials of the British Department of Trade.

References

Adler-Karlsson, G. (1968). *Western Economic Warfare 1947–1967.* Almquist & Wiksell: Stockholm.

Akehurst, M. (1972–1973). Jurisdiction in international law. *British Yearbook of International Law* **XLVI**, 145–257.

Berman, H. J. & Garson, J. R. (1967). United States export controls – past, present and future. *Columbia Law Review* **67**, 868.

Carswell, R. (1982). Economic sanctions and the Iran experience. *Foreign Affairs* **60**, 247–265.

Holloway, J. & Picciotto, S. (1980). Capital, the state and European integration. *Research in Political Economy* **3**, 123–154.

Josephson, M. (1934). *The Robber Barons.* Harcourt Brace Jovanovich: New York.

Kindleberger, C. P. (1969). *American Business Abroad – Six Lectures on Direct Investment.* Yale University Press: New Haven and London.

Lowe, A. V. (1981). Blocking extraterritorial jurisdiction: the British protection of Trading Interests Act, 1980. *American Journal of International Law* **79**, 257.

Lowenfeld, A. F. (1979). Public law in the international arena. *Receuil des cours, Hague Academy of International Law* **163**, 315–445.

Lowenfeld, A. F. (1981). Sovereignty, jurisdiction and reasonableness: a reply to A. V. Lowe. *American Journal of International Law* **75**, 629–638.

Maier, H. G. (1982). Extraterritorial jurisdiction at a crossroads: an intersection between public and private international law. *American Journal of International Law* **76**, 280–320.

Murray, R. (1971). The internationalization of capital and the nation-state. *New Left Review* **67**, 84–109.

U.S. Government (1979). Diplomatic Note No. 56 dated 9 November 1979 from the U.S. Ambassador to British Secretary of State for Foreign Affairs commenting on the Protection of Trading Interests Bill, together with U.K. Diplomatic Note No. 225 of 27 November 1979 in reply.

Vagts Detlev, F. (1982). A turnabout in extraterritoriality. *American Journal of International Law* **76**, 591–549.

Van der Pijl, K. (1979). Class formation at the international level. Reflections on the political economy of Atlantic unity. *Capital & Class* **9**, 1–21.

Warren, B. (1971). The internationalization of capital and the nation-state: a comment. *New Left Review* **68**, 83–88.

Woolcock, S. (1982). *Western Policies on East–West Trade.* Routledge and Kegan Paul for the Royal Institute of International Affairs: London.

PART 3

Introduction
Synthesis: interdependence and
the uniqueness of place

JOHN ALLEN

In the preceding section each of the three chapters focused upon the internal characteristics and relations of a particular aspect of society and drew out their geographical significance. It was apparent, however, that none of the three social aspects – cultural forms, urban economic activity, and the processes of international law – could be conceived in isolation from other aspects of society. Although a series of interrelationships was lightly sketched between cultural, political and economic processes in varying degrees in the analyses, the actual links and connections between the social processes that shape and structure the different aspects of the social world were not developed. This development involves a process of synthesis, a process that takes the results of analysis, the detailed studies of particular aspects of society, and draws out the web of relationships that integrates and binds them to the wider social sphere. Sketched in this manner, the task of synthesis is to construct a more complex geography of social relations from the different geographies of culture, housing, employment, law, and so forth.

By synthesis, however, we wish to convey something more than a simple integration of the various subdivisions of the subject matter of geography. The conception of synthesis we wish to employ is not one of an exhaustive quest for each and every social relationship down to the last detail that comprises the geography of an area. Synthesis addresses the methodological issue raised in the introduction to this book: how the general and the particular are combined in explanation, how the particularity of place is preserved and modified within the generality of social change to produce different outcomes in different places. Clarke's analysis of culture (Chapter 3) caught something of the flavour of the type of synthetic combination that guides the readings in this section when he spoke of cultural change as a product of negotiation and resistance between past and present cultural forms. The distinctiveness of local cultural patterns, he argued, is a product of older cultural strands affecting the shape of new cultural forms. In different places this combination of the old and the new takes different forms depending upon which cultural strands persist, which disappear, and which elements are transformed in the process of change. The result is the geographical unevenness of culture.

The type of explanation Clarke is moving towards, then, is not exhaustive, it is not the geography of all and sundry; a synthesis of elements should be restricted to the interaction between particular social processes, be they economic, political or cultural, and the social characteristics of a specific geographical area. The aim of this type of synthesis is to unravel how broad social processes affect the social structure of certain areas and how, in turn, the economic organization, political character, cultural form and mediated environmental attributes of particular areas shape the actual manner in which the social processes operate. The focus of this mode of synthesis is specific; it attempts to capture the interrelationship of social processes – the links in the social relationships between town and country, region and nation, and between nations – without losing sight of the distinctive history and character which different places bring to bear upon the impact of those processes.

The point can be illustrated by taking the example of urban areas. In the previous section, Ball (Chapter 4) analysed the urban form in terms of its geography of production, population and built environment. He pulled out the detail of each aspect and the links between population movement, location of housing, offices, and so forth and the changing spatial distribution of jobs. Changes in the latter feature, the geography of employment, have received much attention of late as the location of production has moved outwards from the major urban conurbations. This is a general process of capital restructuring that links various cities, towns and regions up and down the country, but what is lost in this general statement is the uneven impact of this process upon particular urban areas. Rising levels of unemployment, for example, are likely to have a varied impact in different urban locations depending upon the organization of social life, political and cultural traditions, and historical experience of an area.

Moreover, the process of restructuring has not led to job loss in all urban areas; some towns have benefited from the changing spatial division of labour. A different geographical pattern of employment and class relations is taking shape. To unravel the impact of the different dimensions of one general process upon particular areas is to understand, first, how the different social elements within an area modify and influence one another to produce a unique place; and second, how the distinctive character of places, towns, cities, fuse with general patterns of social change to produce a geographically uneven social impact.

The two aspects of synthesis are integral to one another and cannot be separated in explanation, although an emphasis upon one or the other aspect leads to a different type of study.

The first aspect is drawn out through an examination of the changing pattern of social relations *within* particular areas, and is illustrated in this section by McDowell and Massey's account of the uneven geography of gender relations (Chapter 7). The second aspect is best seen through a focus upon the interrelationships *between* areas, the social links that bind places together to produce a geographically uneven yet interdependent process of change. Murgatroyd and Urry illustrate this aspect of synthesis in Chapter 6.

In their account of local economic change, they are careful to point out that the effects upon one local economy, Lancaster, of the general restructuring of manufacturing capital over the last decade cannot be deduced from a knowledge of contemporary changes within the nature of industrial capital. The changes in the local pattern of employment are considered as a product of the interaction between, on the one hand, the cultural, political and economic characteristics of the locality, which, in part, stem from the area's historical role in wider successive spatial divisions of labour; and on the other hand, a national and international process of capital restructuring. The ensuing combination of this economic process and the social fabric of Lancaster, which is specific in time, produces, as Murgatroyd and Urry point out, a unique social outcome.

This local pattern of industrial change cannot be accounted for as part of a general industrial decline of the Northwest region of England. The authors point to considerable variations in the past industrial structure of areas in the region, and indeed, to more recent contrasts in the patterns of employment and unemployment within the region. They argue that there is no magical formula which can predict how the process of industrial restructuring will affect different local economies. The effects will be different in different localities, and will be dependent upon the role of the local economy in past and present forms of the spatial division of labour.

In Lancaster, the peculiar characteristics of the locality: the structure of ownership, control, and type of manufacturing and service capital, the paternalistic character of social relations that marks the workforce, the pattern of gender relations outside of the workplace, the local political composition, and physical location of the town, have all contributed towards the contemporary reorganization of the local economy in a new national spatial division of labour.

The thrust of their argument, therefore, is both substantive and conceptual. Conceptual, because they have attempted to synthesize national economic trends with particular historical local circumstances. Both sides of the coin, the general and the unique, are preserved within their conceptual framework. The authors do not offer a local or regional synthesis that is restricted to factors that are to be found only within the boundaries of a particular area; they attempt to integrate a broad process of economic restructuring and reorganization that links a number of geographical localities with the specificity of one local economy. It is this double-sided characteristic of their practice of synthesis that distinguishes it from much of the traditional regional synthesis within geography. Places are not conceived as separate and unrelated; social processes that traverse conventional spatial boundaries such as the region or the nation are integral to an understanding and explanation of the shape and direction of social change at the local or regional level. The interdependence of areas is as important as the uniqueness of areas.

Chapter 7 develops the second aspect of this mode of synthesis in relation to the geography of gender relations: how the uniqueness of areas is constructed and reproduced over time within the context of wider social change. The authors, McDowell and Massey, offer two snapshots in time of four geographical areas to show how the gender relations and women's roles change in different ways in different areas in response, in part, to previous patterns of gender relations in an area, and the wider processes lying behind them.

The structure of the gender relations of an area is conceived by the authors as the product of a combination of historical layers, in which each layer represents a geographically specific articulation of economic and patriarchal relations. Changes in the wider economic organization of society, the successive roles that an area has played within a wider national and international division of labour, are shown to structure and restructure the division of labour between men and women. But there is more to their argument; changes within gender relations are not conceived as a simple effect of economic change – the relation is reciprocal, rather than in one direction. Subsequent changes in the economic base of an area are, to a certain degree, also a reaction to the established pattern of gender relations laid down within previous articulated layers. The two complex layers combine to produce a qualitatively new pattern of gender relations which will vary in its form from area to area.

Conceptualizing the geography of gender relations in this way allows the authors to show how areas characterized by different gender histories, differences in gender roles at home and work, ideologies and attitudes, and so forth, when drawn into similar roles in a new spatial division of labour, throw up distinctive patterns of gender relations. Variety is captured within the interstices of the general.

Chapter 8 offers a synthesis at a different geographical level. Richards and Martin's account of labour migration in the Middle East illustrates the varied impact within Arab nation states of the increase in migrant workers that has accompanied the changing role of Middle Eastern economies in the capitalist world economy. The authors show that the practice of synthesis is not limited to an account of the interactions that occur between, on the one side, the local or regional social structures, and, on the other side, a set of expansive social processes that operate across regional and national spatial boundaries. It is equally possible to talk of syntheses between international forces and national social structures, as it is possible to develop a synthesis between national changes and their local social effect. The difference is merely one of scale, rather than legitimacy. (The categories of geographical area – locality, region, town, city, nation and so forth – do not in themselves carry any explanatory significance: they act merely as a convenient shorthand to describe the geographical extent of social processes and phenomena.)

Our reason, however, for including this article rests upon the authors' detailed account of how geographically located cultural forms have shaped wider economic forces to the extent that the two features are inextricably interrelated. At a general level, the authors point out that economic forces in the form of the oil boom have reinforced a long-standing tradition of labour migration between Arab nations rooted in Muslim culture. A strong demand for labour exists at all levels of skills in the oil exporting countries of Saudi Arabia, Libya, Qatar and the United Arab Emirates. It is at this point that religious culture and economic interests fuse to facilitate a social trend. But the reasons for this demand for migrant labour are not simply economic (a shortage of indigenous labour), they are culturally specific to the area: the low rate of female participation in the workforce and the disdain of the Bedouin population for any kind of manual labour. Both factors derive from historical values that predate the penetration of industrial capital in the Middle East, yet both operate to shape a distinctive process of labour migration that has mushroomed through the integration of Middle Eastern economies in a new international capitalist division of labour. The past has combined with the present to produce a distinctive outcome; the cultural and the economic have qualitatively modified one another in a unique fashion. The one is constitutive of the other.

Richards and Martin go on to show that this rather general synthesis has had more detailed political implications for a number of Middle Eastern states. In their view, the different nationalities of the migrants, the different religious ideologies held, and the different political sympathies of migrant groups produce a complex and varied picture of political interaction. Different countries in the Middle East experience the process of labour migration in different ways, in response to their own political, cultural and economic complexion. Moreover, this structured uneven impact is not limited to the countries that receive migrant labour. The impact of the monetary remittances upon the economies of the sending countries is portrayed as equally varied in response to the different structures of the economies involved. Variety and uniqueness of place are reproduced within the general dynamics of social change.

6

The restructuring of a local economy: the case of Lancaster

LINDA MURGATROYD AND JOHN URRY

1. Introduction

In this chapter we will consider how one particular local economy within Britain has been reorganized over the past twenty or thirty years. We will suggest that this reorganization, reflected in the apparently simple changes in the relative size of manufacturing and service employment, is in fact the product of complex relationships between the underlying 'restructuring' of the various industrial sectors pertinent to the locality. Thus, industrial location and employment changes are not simply the consequence of certain general processes which are merely developed to a lesser or greater extent in any particular local economy. Any such economy must rather be seen as a specific conjuncture, in both time and space, of the particular forms of capitalist and state restructuring within manufacturing and service industries. As Massey argues,

> the social and economic structure of any given local area will be a complex result of the combination of that area's succession of roles within the series of wider, national and international, spatial divisions of labour.
>
> (Massey, 1978, p. 116)

There are three important implications of this 'structural' approach for the analysis of industrial location and employment change. First, the changing forms of the spatial division of labour, especially the shift away from a high degree of regional specialization, derive from new patterns of capital accumulation. In particular, they reflect the internationalization of capitalist accumulation and the development of 'neo-Fordist' methods by which the labour process is controlled.

Second, changes in the location of industry are not to be explained simply in terms of 'economic' or 'political' factors; location is rather to be

Editors' footnote: The original contained very extensive references and substantial data most of which are excluded from this version because of length considerations.

Source: Redundant Spaces in Cities and Regions? Studies in Industrial Decline and Social Change, J. Anderson, S. Duncan and R. Hudson (eds.) (Academic Press, London, 1983), ch. 4, pp. 67–98.

understood in relation to those forms of economic restructuring within and between industrial sectors which are necessitated by the requirements of capital accumulation. However, relations between classes, and other social forces, also significantly affect patterns of economic restructuring, and the latter themselves influence social relations within particular localities to a substantial extent.

Third, problems of uneven development cannot be analysed simply in terms of 'regions' and of regional growth or decline. With the growth of national and international branch circuits there has been a decrease in the degree to which productive systems are centred upon a particular region. This is related to the dispersal of new manufacturing employment on a periphery-centre pattern and to some consequential decline in regional variations in unemployment and economic activity rates between the mid-1960s and the late 1970s. This homogenization among the peripheral regions has also been partly reinforced by the growing concentration of the functions of conception and control within the South East region of the U.K. Similarly, in terms of industrial change, Fothergill and Gudgin (1979, p. 157) conclude that "there are much greater contrasts within any region than between the regions themselves".

In this chapter we shall consider these three points in relation to the de-industrialization of the Lancaster sub-region (which we will refer to simply as 'Lancaster'). This local economy is situated in the north of the North-West Planning Region (north of Preston and Blackburn) and consists of the former urban districts of Lancaster, and Morecambe and Heysham, as well as extensive surrounding rural areas from the Fylde to the Lake District, and from Morecambe Bay in the west to the Pennines in the east. It constitutes a relatively self-contained labour market, with only 7.6% of residents in the district travelling elsewhere to work and 8.7% of those working in it living outside (Census of Population, 1971).

Recently, Lancaster has, like much of the rest of the UK, been 'de-industrialized', with a shift out of manufacturing into either service employment or unemployment. This resulted from a number of underlying processes whose impact in different regions and localities varies greatly. In other words, there is no simple 'de-industrializing' process by which national and subnational economies develop, with one kind of economic activity automatically replacing another as dominant.[...]Therefore to say that a local, regional, or national economy has been 'de-industrialized' is merely a way of *describing* certain shifts in the structure of employment; it does not provide an explanation[...].

In the following we shall show how the 'de-industrialization' of Lancaster has resulted from a combination of processes. Although during the 1950s and early 1960s this subregion was an important site for private manufacturing investment, the national and international reorganization of capital produced a rapid decline in Lancaster's manufacturing base from the 1960s onwards.

We shall focus on three manufacturing industries in greater detail, to show how their particular forms of restructuring had the effect that most of the new employment generated was located outside Lancaster, whether abroad or elsewhere in the UK. We shall also briefly consider the restructuring of the service industries in the locality. We then turn to some of the implications of the process of capitalist restructuring in Lancaster, for local politics and for the state. [...]

2. The restructuring of Lancaster's economy

2.1. Employment in Lancaster

The main changes in the structure of employment in Lancaster, the North-West Planning Region and the UK between 1951 and 1977 can be summarized as follows. Overall, there was a major reduction in manufacturing employment in Lancaster, from around 17000 to 9000 jobs. At the same time, there have been major *increases* in many of the service industries, with employment increased by 5000 over the period, 1951–77. Female employment and unemployment both grew, and while the total population had expanded by 13%, there was an 11% fall in the employed population.

These overall changes did not occur smoothly over the thirty-year period. Manufacturing and service employment both grew considerably, by 9.5% and by 13.2% respectively between 1952 and 1964 (Fulcher *et al.*, 1966). Employment in a number of manufacturing industries grew substantially: textiles by 32.3% (1349 employees), engineering by 56.5% (360), clothing and footwear by 17% (124) and floor-coverings and coated fabrics by 25% (1002). Although the long-established local furniture industry was run down (the main factory closed in 1962), an oil refinery and chemical plant were developed and became major local employers. Certain categories of service employment also showed very substantial rates of employment growth; notably distribution 38%, insurance, banking 35%, professional services 42%, and public administration 19%.

During this period, there was a strong demand for labour, and unemployment was minimal. The labour force increased by over 4000, at a rate about equivalent to the national rate but faster than the average for the North-West region. Much of this increase was accounted for by in-migration, and female employment also increased slowly, but even with these expansions in the labour force, local employers found increasing difficulty in recruiting labour in the early 1960s.

Over the next fifteen years, the situation altered dramatically. Manufacturing employment fell from 16700 in 1961 to 11200 in 1971, dropping to 9000 in 1977 (Department of Employment). By contrast, employment in the service sector expanded steadily from 23000 in 1961 to 26800 in 1971, and rose more sharply during the early 1970s to reach 29400 in 1977, that is, from just over one-half to two-thirds of total employment between 1961 and 1977.

This structural shift has had significant effects upon the differential employment of men and women locally. Between 1951 and 1971, male employment fell by 1200, while female employment rose to 4900. The economic activity rates for men fell from 81.4% to 72% (with 67.4% in employment), while the female rate rose from 31.8% to 36.8% (Census of Population, 1951, 1971). As a result there was an increase in the ratio of female to male workers in the subregion. The general shift towards the service industries is related to the increased feminization of the Lancaster labour force since the proportion of women employed in these industries is high. However, there has also been a shift *within* the service sector towards increased feminization (from 49.1% to 54% between 1971 and 1976; Department of Employment, ERII).

There has also been a considerable growth in declared part-time employment especially during the 1970s, and like elsewhere, there has been a long-term increase in the numbers of registered unemployed.

Up to the mid-1960s, Lancaster was a significant centre for capital accumulation, and this led to a considerable labour shortage. However, from 1965 this became less the case, and the local unemployment rate rose above the national average. The male rate in particular has increased steeply, and a growing gap between the national and local rates has developed.

What explanation can we provide for this shift in the pattern of employment over the post-war period? One obvious explanation of this is that the industries present in Lancaster 1956–60 were those which were about to decline in employment nationally (such as cotton textiles and related indus-tries). However, Fothergill and Gudgin (1979) show that the industrial structure in Lancaster's manufacturing sector in the late fifties was in fact exceedingly favourable in comparison with other areas, and it remained favourable throughout the sixties, when those manufacturing sectors represented in Lancaster expanded in the UK as a whole.

The explanation of the decline in manufacturing employment (3300 between 1959 and 1975) cannot, therefore, be the existing industrial structure. The alternative is that established firms failed to grow, or at least to expand employment, and closed or shrunk instead, while the locality failed to attract mobile employment which was being generated elsewhere in the sixties and early seventies. Conversely, the relatively large expansion of Lancaster's service sector could not have been predicted from the structure of service industries in 1959, but was due rather to the movement of services (such as the University) into the subregion, and the disproportionate growth (or lesser decline) of those already present, such as health services and tourist-related trades. This pattern in Lancaster was directly the converse of that in neighbouring subregions which had a highly unfavourable industrial structure for manufacturing, while that for services was favourable.

This suggests then that the decline of Lancaster's manufacturing base rests with the 'poor performance' of existing capital combined with the inability to attract substantial new plants into the subregion during the 1960s. Hence at a time of very considerable industrial restructuring, with considerable new plant mobility, Lancaster failed to attract such investments and its existing capital became progressively unable to compete with the new plants being established elsewhere.

Fothergill and Gudgin have attempted to explain the overall pattern of employment change in terms of the division between urban conurbations and semi-rural areas. It was in the former, and especially in the large conurbations, that manufacturing employment decline was most marked. This has resulted from the *in situ* contraction of employment and plant closures. And it was in the less industrialized, semi-rural areas that the most substantial relative increases in both manufacturing employment and total employment have been recorded. The Lancaster subregion, in terms of its rural/urban characteristics, fits into the latter category. However, manufacturing employment in Lancaster actually *fell* by 22% between 1959 and 1975.

Lancaster *should* have experienced considerable increases in manufacturing employment, given both its industrial structure in 1959 and the 'performance' of similar less heavily industrialized subregions. But second, identifying a locality in terms of its urban/rural characteristics glosses over a number of highly diverse determinants of employment change. Overall it is more important to identify the place that a particular locality occupies in relationship to the changing spatial division of labour, and hence why in *certain* cases *in situ* expansion or new plants will be developed within smaller less urbanized centres. We will now consider the forms of restructuring of the manufacturing and service sectors in the Lancaster economy.

2.2. *Restructuring in the manufacturing sector*

First, there has been a significant increase in the numbers of establishments and of enterprises. In 1964 there were 58 manufacturing firms in the Lancaster subregion while by 1979 there were at least 135 separate manufacturing establishments employing over ten people, but the increase was largely in the small firms sector. By 1979 there were 109 firms employing between ten and a hundred employees compared with 35 employing below 100 in 1964. Indeed it is interesting to note how successful the subregion has been in establishing and attracting small manufacturing firms. Partly this may have resulted from the efforts of the City Council to encourage the establishment and growth of such firms, not simply because this is the only alternative, but also because of the perceived problems caused by dependence on 'externally controlled' capital, graphically highlighted when the Lansil works owned by British Celanese (part of Courtaulds) closed in 1980. This episode also shows very well the limitations of the popular 'small firm solution' to industrial decline. This one closure caused more jobs to be lost than had been created by small firms during the *whole* of the 1970s.

Since the late 1950s, a high proportion of manufacturing employment has been in firms which were controlled from outside the immediate area; and there has been a slight increase in this proportion. Subsidiaries or branches of large companies tended to be the larger establishments. A considerable number of locally based firms were taken over by, or merged with, other companies over this period.[...] This dependence of the local economy upon a small number of large externally controlled firms was noted by the local planning department:

...it is significant that a very high proportion of closures and redundancies declared during recent periods of recession have been in firms who are under external control, i.e. 'pruning the branches to encourage growth'.

(Lancaster City Council: 1977, Appendix IIIc)

The 'branches' are located in Lancaster.

It is also important to note here that *all* the large manufacturing firms in Lancaster are now externally controlled, while independent ownership is much more characteristic of the very small firms. Moreover, hardly any of the large plants were established by major multinationals; rather they were locally-owned firms which were *acquired* by (or merged with) large companies based elsewhere. In other words, external takeovers have been far more important for the employment structure of Lancaster than have patterns of branch-plant migration (except in the 1940s and 1950s with the establishment of a fertilizer and ammonia plant by ICI and a refinery by Shell). In general, Lancaster has not developed as a branch-plant economy in the sense of branch-plants moving in; rather, existing plants have become branches.[....]

We will now consider three forms of restructuring of industrial sectors which will generate different spatial patterns of employment decline (see Massey and Meegan, 1982). They are 'intensification', 'investment and technical change', and 'rationalization'. *Intensification* is the process of increasing the productivity of labour with little, if any, loss of capacity and no investment in new forms of production. Such a reorganization of production will generally entail less change in the distribution of employment than any of the other forms of reorganization. In the second type, *investment and technical change*, there is heavy capital investment in new forms of production and as a result considerable job-loss, often highly unequally distributed among job-types and skills. The third form, *rationalization*, produces closure of capacity without new investment or change in technique.[...]

The restructuring of the British economy in this way has had important spatial implications, and Lancaster seems to have been one of the casualties. Lancaster's manufacturing employment was concentrated in industries in which 'investment and technical change', or 'rationalization' were to occur, rather than 'intensification'. To demonstrate this in detail is beyond the scope of this chapter; instead, we will discuss the broad trends in the three principal

118   *Linda Murgatroyd and John Urry*

manufacturing industries. In June, 1971, linoleum, plastics and floor-coverings, made-made fibres and fertilizers employed 2502, 2015, and 911 people respectively, and between them accounted for 46% of all manufacturing employment in Lancaster. By 1977, employment in these industries had fallen by 41%, 66% and 32% compared with their 1971 levels. What, then, were the forms in which production in these industries was reorganized, to produce such a fall in their employment in Lancaster?

(1) Floorcoverings, linoleum, plastics, etc.

[....] During the 1950s this was a relatively buoyant industrial sector with a 25% increase in the local employed labour force. However, during the 1960s there was a sharp decline in employment within this sector. The local results were that:

...the major company since the 1860s, the Mills [Williamsons], were forced into a defensive merger with a major rival; the merger led to considerable rationalisation, the disposal of surplus assets and the consolidation of both administration and production at the headquarters of the former rival [Nairns], in a government development area [Kirkcaldy].

(Martin and Freyer, 1973, p. 168: names in brackets added).

In fact, all floor-covering production was transferred to Kirkcaldy and the Lancaster plant mainly concentrated on PVC wall-coverings instead.

Linoleum manufacture had long been established in Lancaster, having developed from sail-cloth and oil-cloth production in the nineteenth century, and much of the equipment was therefore outdated. The market for these products deteriorated sharply with the development of plastics, and later of cheap carpets based on man-made fibres. Local firms did diversify into PVC sheeting and various kinds of coated fabrics and wall-coverings, but much of this was achieved by the adaptation of old machinery rather than by substantial new capital investments and the use of new forms of production technology. The main production change within this industry has been the decline in linoleum and the development of plastic floor coverings. This resulted from a combination of two processes: rationalization and the almost complete disappearance of manufacturing capacity in linoleum, and technical change and investment within plastic floor and wall-coverings. This has had dramatic employment effects.

Overall, the rate of decline in this sector has been far faster in Lancaster than in the UK as a whole. Nationally, restructuring (including closures) has been accompanied by some expansion, but this has not occurred in Lancaster.

(2) Fertilizers

In contrast to the floor-coverings industry, there have been enormous increases in output, capital expenditure and productivity in this industry nationally. Both output and productivity increased ten times over the period 1963–78, while capital investment increased nine times between 1968 and 1978 (all in money terms). The increases in productivity in fertilizers were greater than for any other branch of the chemicals industry between 1970 and 1975, but the capital investment was concentrated in the development of new, very large, low cost manufacturing plant: by the late 1960s there were six major plants in the UK. As a consequence there was a 20% reduction in the numbers of both establishments and enterprises between 1963 and 1978. Moreover, these enormous increases in output were achieved with little or no increase in total employment in the fertilizer industry. The fertilizer industry is a very good example of 'jobless growth' where the process of restructuring took the form of 'investment and technical change'. What then were the consequences for the Lancaster economy?

[Employment in fertilizer production in Lancaster fell from 2285 in 1952 to 624 in 1977.] The workforce fell in the main local plant (ICI at Heysham) for two main reasons. First, ammonia production was abandoned in 1977, and concentrated in the larger plants, especially in the North-East. And second, the fertilizer made at Heysham (Nitrogel) could not compete with the new fertilizer (Nitran) which had been developed by ICI in the mid-1960s. The Heysham plant is disadvantaged in that its capacity is too small (500 tons per day compared to 1500/2000 tons at more modern plants) and because none of its main production capacity dates from later than 1962. Again we see how capital accumulation in manufacturing industry has not led to reinvestment in Lancaster, and that this decline results from the development *elsewhere* of newer, cheaper manufacturing capacity.

(3) Man-made fibres

Production of man-made fibres began in Lancaster in 1928 with the establishment of Cellulose Acetate Silk Co. Ltd (later known as Lansils), in an area of considerable national expansion in the production of rayons. By 1929 there were 32 national competitors; yet in the next 30 years one firm, Courtaulds, came to dominate rayon production. However by the early 1950s the market for rayon was being eroded by the development of synthetic fibres (especially nylon and later polyesters) and of improved cotton. Courtaulds itself diversified and came to dominate production in all sections of the textile industry, except weaving which remained rather fragmented, and in 1973–4 it acquired Lansils in Lancaster.

Two points are important here. First, man-made fibres constitute a sector in which investment and technical change has been particularly marked, especially in the period up to 1970. Courtaulds has been able to enlarge its monopoly position by moving into synthetics, and by vertical integration. Second, the effect of this in Lancaster has been the closure of Lansils in 1980, after a long period of decline. Courtaulds justify closure with reference to annual losses of £91000, but more fundamentally this resulted from failure to update or replace machinery since first installation in 1952. It is true that the reorganization of the textile industry as a whole: mergers, acquisitions, technical changes, and new plants in development areas and abroad, produced new accumulation away from the traditional Lancashire textile towns. But we might have expected Lancaster to have been protected from some of the worst consequences of this process because of its involvement in the production of man-made fibres rather than cotton. However, this did not occur. Accumulation in the 1950s was followed by disinvestment in Lancaster in the 1960s and 1970s. The total number of textile workers in the travel-to-work-area has declined from 5500 in 1964 to 1800 in 1977.

To summarize, Lancaster benefited from accumulation in manufacturing industry in the 1950s, and by the end of the decade had high representation in a number of growing industrial sectors. However, these were sectors that were to experience 'investment and technical change', and 'rationalization', particularly because of the increased centralization of ownership. New plants were established elsewhere, while existing plants based on earlier technologies shed labour. Local branches of multi-plant companies were run down, as these companies restructured their production away from Lancaster. While the number of small manufacturing enterprises swelled during the later period, these did not provide sufficient employment to offset the decline in the larger establishments. It is interesting to note that of those industries in which there was much less job-loss in Lancaster, two appear as 'intensifiers' between 1968 and 1973, according to Massey and Meegan (1982, ch. 3). These are textile finishing, whose employment in Lancaster fell from 432 to 278 between 1971–7, and footwear whose employment fell from 549 to 414.

2.3. The reorganization of services

We will now consider how restructuring in service sector industries has affected Lancaster. First of all, however, it is necessary to distinguish between service *industries* and service *occupations*. The former are those industries with a product which is classified as a service rather than a tangible good, and include transport and communications, distribution, professional services, and so on. The latter, service occupations, are like service industries in that they are normally defined in contrast with the *physical* production of commodities, and include managers, professionals, clerical and sales workers, health and educational workers and so on. The most important implication of this distinction is that there are many workers within manufacturing industry who are located within the service occupations. Crum and Gudgin (1978, 5) estimate that by 1971, 34.6% of manufacturing workers in the UK were what they term 'non-productive' in this sense. So another way of expressing the 'de-industrialization' thesis is to point out that the absolute number of 'direct production workers' had fallen from 6.5 m. in the late 1950s to 5.2 m. in 1975.

Partly this reflects the changing industrial structure, so that roughly speaking the later an industry develops, the higher the proportion of non-productive workers employed within it. It also reflects the socialization of non-productive labour within firms, and the differentiation of the functions of management between a large number of agents as conception is increasingly separated from execution.

We can also identify some distinctive factors which affect the labour process within service work. First, labour-power has to be expended more closely to where the consumer demands it, and this has implications for the spatial structuring of such industry. It is also more difficult to standardize the product and hence to fragment the labour process as in manufacturing industry. There is also some degree of control maintained by many service workers over the nature of their work. This is true even within distribution, clerical and secretarial work, and is particularly marked in the case of work involving personal contacts with 'clients' (e.g. in the Health Service).

In the service sector, capital (and the state) tend to economize on labour costs, not principally through direct increases in productivity (though this of course happens) but rather through the employment of sectors of the labour force which can be employed at less than the average wage for white males. In most of the major capitalist economies, there have been much larger increases in the employment of women than of men in the growing service sector. (A significant exception to this is West Germany, where large numbers of 'guestworkers' have been employed.) In the UK, women are now five times more likely to be employed in service than manufacturing industry. This development is connected to some de-skilling of service employment. We should note that such variations in women's participation rates will have significant implications for the cost of reproducing labour power, the size of local labour reserves, and the levels of organization and politicization of the labour force; all of which may in turn affect patterns of industrial restructuring (see Chapter 7).

What then has happened to the service industries in Lancaster? Between 1951 and 1977, the numbers employed in them increased by 21%, as compared with a 23% increase nationally. This constituted an increase of over 5000, at the same time that total employment had fallen by nearly 6000. However, this overall expansion conceals a number of divergent trends.

Between 1960 and 1979 employment in transport and communication declined by over one third, while that in professional and scientific services almost trebled (up 180%). Other service industries maintained fairly steady levels of employment. In the relatively large size of the transport sector, and the small numbers employed in financial services and government administration, Lancaster was fairly typical of the North-West as a whole. During the 1950s Lancaster had a lower proportion of people employed in professional and scientific services than was the national average (7.5% compared with 7.9% nationally), and in this also it resembled the average for the North-West Region. However, during the 1960s and 1970s, the expansion of the sector resulted in strong local concentration of employment in these services. Of the employed labour force in Lancaster, 20% were in this sector in 1977, compared with 16% nationally.

Many of these shifts result not from changes in local markets or other indigenous factors, but from decisions taken at a national level, mainly concerning changes in the railway, education and health system. As in the case of manufacturing industry, the domination of the transport and the professional and scientific services by organizations which extended beyond the boundaries of Lancaster resulted in reorganization which affected the locality to a disproportionate extent. While the 'market' for many of these services is local, in the sense that health and education authorities cater for those living within their boundaries, there has also been a concentration in Lancaster of specialized areas of health care (e.g. mental hospitals, geriatrics), and of higher education which serve a population far wider than that permanently living in the travel-to-work area. Both employees and clients in these services are geographically mobile and moved into the area in order to take up the jobs or services available in Lancaster.

The tourist industry is the other major employer in the area. Here again, there is a net invisible export from Lancaster via the geographic mobility of the clientele, a large proportion of whom travel from other parts of the North-West. 'Miscellaneous services' and 'Distribution' are the two industries most closely connected with tourism.

In 1951 'miscellaneous services' (which includes cinemas and theatre, sport and recreation, betting and gambling, hotels, restaurants, public houses, clubs, etc....) was by far the largest service industry in the area, accounting for 13.5% of all local employment; double the proportion nationally. There were fluctuations in the level of employment in this sector, but by 1977 the net increase since 1951 was minimal, despite their increased *share* of employment locally.

Similarly, distributive trades maintained a steady level of employment over the period, apart from a drop in the mid-1960s attributable to selective employment tax. The steady level of employment in this sector resulted from two opposing forces; nationally, retailing employment declined due to a shift towards large-scale outlets, but the population serviced by retailers in the Lancaster district increased, due to a high rate of in-migration of professional and retired families. The local multiplier effects of the expanded health and education services, more than made up for the slight decline in tourist-related employment over the period.

Although little detailed information is available concerning changing ownership patterns in the local service sector, it is clear that there is a trend towards external ownership of the large enterprises in this sector, just like in manufacturing. There is also an increased local dependence on public sector employment in educational and health services. Not only has an increasing proportion of the local population come to depend directly on the state sector for employment, but also a great deal of employment in other services depends on the incomes generated by these sectors. As manufacturing employment declines, the economy of Lancaster has become increasingly dependent on the level and direction of state expenditure.

2.4. Summary

In conclusion to this section we should note how there has been a substantial shift in the character of this locality over the past thirty years. In 1950 the local economy was dominated by a small number of private manufacturing employers, who were involved in numerous commodity and interpersonal linkages with the locality and with the surrounding textile-based region. In 1980, the state is the dominant employer, and the fortunes of the small private employers depend upon the expansion or contraction of state expenditure, principally within the service sector. Relatively few large manufacturing establishments remain, and they have limited linkages with other locally based firms.[...] In the next section we will briefly discuss some of the local and policy factors involved in these changes and how these changes in turn affected the forms of economic and political struggle within Lancaster.

3. State policies, politics and struggle

There are three significant issues to deal with here. First, why was Lancaster unable to attract the mobile new employment that was generated in those industries undergoing technical change and making new investments? Second, what have been the characteristic features of the economic and social relations in Lancaster that have influenced this failure? And third what have been the consequences of the changes in the ownership and the structure of employment upon local struggles?

3.1. Regional and local industrial policies

On the first question we may begin by noting that Lancaster is part of the North-West Region, and this region has performed very badly in employment terms over the recent period. Stillwell maintains that the North-West was among "the least attractive regions in which to locate industry" (1968, p. 10). This meant that other regions attracted the mobile plants which made a major difference in employment terms. This produced cumulative disadvantages for those less-favoured areas, as the age of the region's capital stock got progressively older and less competitive with the new plant being established elsewhere.

Lancaster should have been partly protected from this effect, given its relative expansion in the 1940s and 1950s. However, this was not sustained, partly because the North-West Region has not constituted an important force politically (in comparison with Scotland or Wales, for example), and partly because Lancaster has had little chance of making effective representation on its own (although it has maintained Intermediate Area Status). The labour movement never developed a strong regional base here, by contrast, for example, with South Wales or the North-East of England. One crude indicator of this is given by the fact that the proportion of people voting Labour in Lancashire has generally been lower than in corresponding regions. (In 1974, 50.1% in Lancashire voted Labour, compared with 59.4% in the North-East.) Lancaster itself comprises two constituencies, yet the first Labour M.P. was not elected until 1966, although the proportion of manual workers in the labour force was over 50% until the 1960s.

The local state has concentrated upon two policies: first, to attract new service employment within the public sector, hence the university and expanded hospital services; and second, to develop small manufacturing firms. The latter policy was introduced in the early 1960s and it was consolidated in the late 1960s and 1970s. This has not been substantially changed, even when unemployment began to rise. Land and technical and financial assistance were made available and a 'seed-bed' experiment was set up to help very small firms to become established. Preference for these facilities was actively given to those small firms, with 'high quality' products, in technologically-based industries. A substantial number of such firms were successfully brought to (or started in) Lancaster, using the facilities provided by the council and the university through *Enterprise Lancaster*, and also helped by the *Small Firms Club* initiated by the city council. Large-scale manufacturing investments were less strongly encouraged by the local council throughout the sixties and early seventies, and Lancaster's designation as an Intermediate Area during the era of Regional Policy after 1972 did not facilitate the attraction of large-scale capital during a period of massive industrial restructuring. In general, regional policy has benefited those areas which experienced full special Development Area status during the central period of regional policy (*c.* 1965–75) at the expense of other regions. However, the impact of regional policy should not be over-estimated, since the period in which it was particularly developed was also that in which the most substantial restructuring of capitalist industry took place.

3.2. *Restructuring class relations and local politics*

We have already noted that the Lancaster subregion has not been an area with a strong labour movement, but contrary to right-wing commentators, this did not result in the attraction of large flows of capital to the subregion, so that they might profit from the quiescent (or 'realistic') labour force. It has been argued that the quiescence of the Lancaster labour force has resulted from the traditional paternalist character of social relations both within work-places and between the local firms and the city. Norris (1978, pp. 471–2) defines paternalism (of the sort once found in Lancaster) as existing where inequalities of economic and political power are "stabilized through the legitimating ideology of traditionalism". He suggests that there are four components to such an ideology: "gentlemanly ethic", "personal dependence", "localism", and a "gift relationship".

While traditional forms of paternalism clearly no longer existed by the 1960s and 1970s, vestiges of these practices are indicated by the responses of the local labour movement to the mass redundancies and plant closures which have characterized 1980 and 1981. These events elicited fatalistic responses from the labour force, and the only negotiations that took place were about the terms of redundancies, their necessity being accepted from the start. Despite an understanding that the closure of Lansil plant (1980) by Courtaulds was caused, not simply by low company profits, but by the company's worldwide restructuring plans, the workers from Lansil's eventually appeared grateful to accept the minimal redundancy payments made by Courtaulds (source: interview with shop steward). This kind of response was in sharp contrast to the occupation simultaneously taking place at Gardners in Eccles, only 40 miles away, and to similar protests elsewhere. Such action was not seriously considered in Lancaster by any of the workforces affected by redundancies, rather notions of the "gentlemanly ethic" and the "gift-relationship" prevailed.

The main active response was also characterized by attitudes associated with paternalism, namely localism and personal dependence. *The Save Lancaster Campaign* was established by the Trades Council late in 1980, in the wake of several redundancy announcements, and the emphasis of the campaign was firmly on the locality rather than on class politics. This campaign did not gain much active support even at this time, and it withered away after a few weeks. Most of those affected preferred either to depend on the provisions of the state and the efforts of the city council, or to find individualistic solutions to unemployment. It may well be that the blossoming of small businesses during the 1970s was an accommodating local response to the decline in employment in older manufacturing firms; in addition, the existence of a large (traditional) petit bourgeoisie (the self-employed) probably undermined collectivist protests.[...]

To some extent it appears that the labour movement had more involvement in trying to 'save' the city's industry than had most other groupings. However, its efforts to preserve capitalist manufacturing activity in the locality clearly deflected labourist struggles away from the traditional issues of the wage-relation or the forms of capitalist control, concentrating them instead on presenting the city as a suitable site for capitalist accumulation. To some extent the labour movement has also put its weight behind the 'small firms strategy', being apparently unaware of the deficiencies of such strategy.

To the extent that social relations in Lancaster were once characterized by 'corporate paternalism', this may have been replaced by a kind of state paternalism. Clearly the local labour force (and indeed the local economy) is dependent to a crucial degree upon the state, and it has responded gratefully when announced cutbacks are less than they might have been. A good recent example has been the attitudes of gratitude and deference exhibited locally when the Manpower Services Commission created considerable temporary employment in the area. However, it may be more appropriate to regard the responses to heightened 'external control' of the local economy as characterized more by *fatalism* than by paternalism.

A number of other developments in local politics can also be mentioned. In particular, various 'oppositional fragments' have emerged, which have been concerned with struggles in the area of consumption as well as production. Such issues as ecology, sexual politics, transport, leisure and the arts have grown in importance locally, as the service sector has come to dominate the area's industry; many of those active in such 'fragments' being either employed in the service sector, or unemployed. The restructuring of the local economy has therefore involved not only the undermining of [already weakly developed] traditional forms of class conflict, but also the development of new struggles and a restructuring of local [political activity].

4. Conclusion

We have thus tried to show how the Lancaster economy has been transformed as a consequence of its location within the changing forms of the spatial division of labour. During the period of post-war reconstruction, based on the expansion of national capital, Lancaster benefited and developed in a number of growing industrial sectors. But with the industrial restructuring of the 1960s and early 70s, a new, in part international, spatial division of labour developed from which Lancaster failed to benefit. Indeed since its fixed capital was of a previous vintage, the effect of the new round of accumulation was to undermine those industries established within the previous round. The main expansion was in state service employment, and partly in private service employment. There was an increasing gap between the relatively skilled employment available in the service sector (especially that of the state) and that de-skilled employment available in the private manufacturing sector. The political composition of Lancaster interestingly reflects this particular combination of forms under which the local economy has been restructured.

Acknowledgements

We are very grateful for the assistance, advice and encouragement of other members of the Lancaster Regionalism Group. We are also indebted to the Department of Employment, to Mr R. H. Kelsall of Enterprise Lancaster, and others, for providing us with information, and to Mr M. Lee for assistance in processing it. This work was financed by the Human Geography Committee of the S.S.R.C., 1980–1.

References

Crum, R. E. and Gudgin, G. (1977). "Non-production activities in the UK manufacturing industry", *Brussels Commission of the European Community Regional Policy Series*, **3**.

Fothergill, S. and Gudgin, G. (1979). "Regional employment change: a subregional explanation", *Progress and Planning*, **12**, 155–220.

Fulcher, M. N., Rhodes, J. and Taylor, J. (1966). "The economy of the Lancaster sub-region", University of Lancaster Economics Department, Occasional Paper 10.

Lancaster City Council. (1977). "Industrial strategy for Lancaster", Lancaster Town Hall, unpublished.

Martin, R. and Freyer, B. (1973). *Redundancy and Paternalist Capitalism*, Allen and Unwin, London.

Massey, D. (1978). "Regionalism: some current issues", *Capital and Class*, **6**, 106–125.

Massey, D. and Meegan R. (1978). "Industrial restructuring versus the cities", *Urban Studies*, **15**, 3.

Massey, D. and Meegan, R. (1982). *The Anatomy of Job Loss: The How, Why and Where of Employment Decline*, Methuen, London.

Norris, G. (1978). "Industrial paternalism, capitalism and local labour markets", *Sociology*, **12**, 469–89.

7

A woman's place?

LINDA McDOWELL AND DOREEN MASSEY

The nineteenth century saw the expansion of capitalist relations of production in Britain. It was a geographically uneven and differentiated process, and the resulting economic differences between regions are well known: the rise of the coalfields, of the textile areas, the dramatic social and economic changes in the organization of agriculture, and so forth. Each was both a reflection of and a basis for the period of dominance which the UK economy enjoyed within the nineteenth-century international division of labour. In this wider spatial division of labour, in other words, different regions of Britain played different roles, and their economic and employment structures in consequence also developed along different paths.

But the spread of capitalist relations of production was also accompanied by other changes. In particular it disrupted the existing relations between women and men. The old patriarchal form of domestic production was torn apart, the established pattern of relations between the sexes was thrown into question. This, too, was a process which varied in its extent and in its nature between parts of the country, and one of the crucial influences on this variation was the nature of the emerging economic structures. In each of these different areas 'capitalism' and 'patriarchy' were articulated together, accommodated themselves to each other, in different ways.

It is this process that we wish to examine here. Schematically, what we are arguing is that the contrasting forms of economic development in different parts of the country presented distinct conditions for the maintenance of male dominance. *Extremely* schematically, capitalism presented patriarchy with different challenges in different parts of the country. The question was in what ways the terms of male dominance would be reformulated within these changed conditions. Further, this process of accommodation between capitalism and patriarchy produced a different synthesis of the two in different places. It was a synthesis which was clearly visible in the nature of gender relations, and in the lives of women.

This issue of the synthesis of aspects of society within different places is what we examine in the following four subsections of this chapter. What we are interested in, in other words, is one complex in that whole constellation of factors which go to make up the uniqueness of place.

We have chosen four areas to look at. They are places where not only different 'industries' in the sectoral sense, but also different social forms of production, dominated: coal mining in the north-east of England, the factory work of the cotton towns, the sweated labour of inner London, and the agricultural gang-work of the Fens. In one chapter we cannot do justice to the complexity of the syntheses which were established in these very different areas. All we attempt is to illustrate our argument by highlighting the most significant lines of contrast.

Since the construction of that nineteenth-century mosaic of differences all these regions have undergone further changes. In the second group of sections we leap ahead to the last decades of the twentieth century and ask 'where are they now?'. What is clear is that, in spite of all the major national changes which might have been expected to iron out the contrasts, the areas, in terms of gender relations and the lives of women, are still distinct. But they are distinct in different ways now. Each is still unique, though each has changed. In this later section we focus on two threads in this reproduction and transformation of uniqueness. First, there have been different changes in the economic structure of the areas. They have been incorporated in different ways into the new, wider spatial division of labour, indeed the new international division of labour. The national processes of change in the UK economy, in other words, have not operated in the same way in each of the areas. The new layers of economic activity, or inactivity, which have been superimposed on the old are, just as was the old, different in different places. Second, however, the impact of the more recent changes has itself been moulded by the different existing conditions, the accumulated inheritance of the past, to produce distinct resulting combinations. 'The local' has had its impact on the operation of 'the national'.

The nineteenth century

Coal is our life: whose life?

Danger and drudgery; male solidarity and female oppression – this sums up life in the colliery villages of Co. Durham during much of the nineteenth century. Here the separation of men and women's lives was virtually total: men were the breadwinners, women the domestic labourers, though hardly the 'angels of the house' that featured so large in the middle class Victorian's idealization of women. The coal mining areas of Durham provide a clear example of how changes in the economic organization of Victorian England interacted with a particular view of women's place to produce a rigidly hierarchial and patriarchal society. These villages were dominated by the pits and by the mine owners. Virtually all the men earned their livelihood in the mines and the mines were an almost exclusively male preserve, once women's labour was forbidden from the middle of the century. Men were the industrial proletariat selling their labour power to a monopoly employer, who also owned the home. Mining was a dirty, dangerous and hazardous job. Daily, men risked their lives in appalling conditions. The shared risks contributed to a particular form of male solidarity, and the endowment of their manual labour itself with the attributes of masculinity and virility. The shared dangers at work led to shared interests between men outside work: a shared pit language, shared clubs and pubs, a shared interest in rugby. Women's banishment from the male world of work was thus compounded by their exclusion from the local political and social life.

Jobs for women in these areas were few. Domestic service for the younger girls; for married women poorly paid and haphazard work such as laundry, decorating or child care. But most of the families were in the same position: there was little cash to spare for this type of service in families often depending on a single source of male wages. For miners' wives almost without exception, and for many of their daughters, unpaid work in the home was the only and time-consuming option. And here the unequal economic and social relationships between men and women imposed by the social organization of mining increased the subordinate position of women. A miner's work resulted in enormous domestic burdens for his wife and family. Underground work was filthy and this was long before the installation of pithead showers and protective clothing. Working clothes had to be boiled in coppers over the fire which had to heat all the hot water for washing clothes, people and floors. Shift work for the men increased women's domestic work: clothes had to be washed, backs scrubbed and hot meals prepared at all times of the day and night:

'I go to bed only on Saturday nights', said a miner's wife; 'my husband and our three sons are all in different shifts, and one or other of them is leaving or entering the house and requiring a meal every three hours of the twenty four.'

(Webb, 1921, pp. 71–2)

An extreme example, perhaps, but not exceptional.

These Durham miners, themselves oppressed at work, were often tyrants in their own home, dominating their wives in an often oppressive and bullying fashion. They seem to have "reacted to [their own] exploitation by fighting not as a class against capitalism, but as a gender group against women – or rather within a framework of sex solidarity against a specific woman chosen and caged for this express purpose" (Frankenberg, 1976, p. 40). Men were the masters at home. Here is a Durham man, who himself went down the pits in the 1920s, describing his father:

He was a selfish man. If there was three scones he'd want the biggest one. He'd sit at the table with his knife and fork on the table before the meal was even prepared...Nobody would get the newspaper till he had read it.

(Strong Words Collective, 1977, pp. 11–12)

Thus gender relations took a particular form in these colliery villages.

National ideologies and local conditions worked together to produce a unique set of patriarchal relations based on the extreme separation of men's and women's lives. Masculine supremacy, male predominance in every area of economic and social life became an established, and almost unchallenged, fact. Patriarchal power in this part of the country remained hardly disturbed until the middle of the next century.

Cotton towns: the home turned upside down?

The images of homemaker and breadwinner are of course national ones, common to the whole of capitalist Britain, and not just to coalfield areas. But they were more extreme in these regions, and they took a particular form; there were differences between the coalfields and others parts of the country.

The cotton towns of the north-west of England are probably the best-known example from, as it were, the other end of the spectrum, and a major element in this has been the long history of paid labour outside the home for women. It is often forgotten to what extent women were the first labour-force of factory-based, industrial capitalism. "In this sense, modern industry was a direct challenge to the traditional sexual division of labour in social production" (Alexander, 1982, p. 41). And it was in the cotton industry around Manchester that the challenge was first laid down.

Maintaining patriarchal relations in such a situation was (and has been) a different and in many ways a more difficult job than in Durham. The challenge was nonetheless taken up. Indeed spinning, which had in the domestic organization of the textile industry been done by women, was taken over by men. Work on the mule came to be classified as 'heavy', as, consequently, to be done by men, and (also consequently) as skilled (Hall, 1982). The maintenance of male prerogative in the face of threats from women's employment, was conscious and was organized:

The mule spinners did not leave their dominance to chance...At their meeting in the Isle of Man in 1829 the spinners stipulated 'that no person be learned or allowed to spin except the son, brother, or orphan nephew of spinners'. Those women spinners who had managed to maintain their position were advised to form their own union. From then on the entry to the trade was very tightly controlled and the days of the female spinners were indeed numbered.

(Hall, 1982, p. 22)

But if men won in spinning, they lost (in those terms) in weaving. The introduction of the power loom was crucial. With it, the factory system took over from the handloom weavers, and in the factories it was mainly women and children who were employed. This did present a real challenge:

The men who had been at the heads of productive households were unemployed or deriving a pittance from their work whilst their wives and children were driven out to the factories.

(Hall, 1982, p. 24)

Nor was 'the problem' confined to weavers. For the fact that in some towns a significant number of married women went out to work weaving meant that further jobs were created for other women, doing for money aspects of domestic labour (washing and sewing, for example) that would otherwise have been done for nothing by the women weavers. Further, the shortage of employment for men, and low wages, provided another incentive for women to earn a wage for themselves (Anderson, 1971).

The situation caused moral outrage among the Victorian middle classes and presented serious competition to working-class men. There was "what has been described as 'coincidence of interests' between philanthropists, the state – representing the collective interests of capital – and the male working class who were represented by the trade union movement and Chartism – which cooperated to reduce female and child labour and to limit the length of the working day" (Hall, 1982, p. 25). In the same way, it was at national level that arguments about 'the family wage' came to be developed and refined as a further means of subordinating women's paid labour (for pin money) to that of men's (to support a family). The transformation from domestic to factory production, a transformation which took place first in the cotton towns,

provoked, as can be seen, a period of transition and re-accommodation in the sexual division of labour. The break-up of the family economy, with the threat this could present to the male head of household, who was already faced with a loss of control over his own labour, demanded a re-assertion of male authority.

(Hall, 1982, p. 27)

Yet in spite of that reassertion, the distinctiveness of the cotton areas continued. There were more women in paid work, and particularly in relatively skilled paid work, in the textile industry and in this part of the country, than elsewhere:

In many cases the family is not wholly dissolved by the employment of the wife, but turned upside down. The wife supports the family, the husband sits at home, tends the children, sweeps the room and cooks. This case happens very frequently: in Manchester alone, many hundred such men could be cited, condemned to domestic occupations. It is easy to imagine the wrath aroused among the working-men by this reversal of all relations within the family, while the other social conditions remain unchanged.

(Engels, 1969 edn, p. 173)

This tradition of waged-labour for Lancashire women, more developed than in other parts of the country, has lasted. Of the early twentieth century, Liddington writes "Why did so many Lancashire women go out to work? By the turn of the century economic factors had become further reinforced by three generations of social conventions. It became almost unthinkable for women *not* to work" (1979, pp. 98–9).

And this tradition in its turn had wider effects. Lancashire women joined trade unions on a scale unknown elsewhere in the country: "union membership was accepted as part of normal female behaviour in the cotton towns" (Liddington, 1979, p. 99). In the nineteenth century the independent mill-girls were renowned for their cheekiness; of the women of the turn-of-the-century cotton towns, Liddington writes: "Lancashire women, trade unionists on a massive scale unmatched elsewhere, were organized, independent and proud" (1979, p. 99). And it was from this base of organized working women that arose the local suffrage campaign of the early twentieth century. "Lancashire must occupy a special place in the minds of feminist historians. The radical suffragists sprang from an industrial culture which enabled them to organize a widespread political campaign for working women like themselves" (p. 98).

The radical suffragists mixed working-class and feminist politics in a way which challenged both middle-class suffragettes and working-class men. In the end, though, it was precisely their uniqueness which left them isolated – their uniqueness as radical trade unionists *and* women, and, ironically, their highly regionalized base:

The radical suffragists failed in the end to achieve the political impact they sought. The reforms for which they campaigned – of which the most important was the parliamentary vote – demanded the backing of the national legislature at Westminster. Thousands of working women in the Lancashire cotton towns supported their campaign, and cotton workers represented five out of six of all women trade union members. No other group of women workers could match their level of organization, their (relatively) high wages and the confidence they had in their own status as skilled workers. Their strength, however, was regional rather than national, and when they tried to apply their tactics to working-class women elsewhere or to the national political arena, they met with little success. Ultimately the radical suffragists' localised strength proved to be a long-term weakness.

(Liddington, 1979, p. 110)

The rag-trade in Hackney: a suitable job for a woman?

But there were other industries in other parts of the country where women were equally involved in paid labour, where conditions were as bad as in the cotton mills, yet where at this period not a murmur was raised against their employment. One such area was Hackney, dominated by industries where sweated labour was the main form of labour-organization.

What was different about this form of wage relation for women from men's point of view? What was so threatening about women working? Hall (1982) enumerates a number of threads to the threat. The first was that labour was now *waged* labour. Women with a wage of their own had a degree of potentially unsettling financial independence. But Lancashire textiles and the London sweated trades had this in common. The thing that distinguished them was the spatial separation of home and workplace. The dominant form of organization of the labour-process in the London sweated trades was homeworking. The waged-labour was carried out in the home; in Lancashire, birthplace of the factory-system, waged-labour by now meant leaving the house and going to the mill. It wasn't so much 'work' as 'going out to' work which was the threat to the patriarchal order. And this in two ways: it threatened the ability of women adequately to perform their domestic role as homemaker for men and children, and it gave them an entry into public life, mixed company, a life not defined by family and husband.

It was, then, a change in the social *and the spatial* organization of work which was crucial. And that change mattered to women as well as men. Lancashire women did get out of the home. The effects of homeworking *are* different: the worker remains confined to the privatized space of the home, and individualized, isolated from other workers. Unionization of women in cotton textiles has always been far higher than amongst the homeworking women in London.

Nor was this all. For the *nature* of the job also mattered in terms of its potential impact on gender relations:

Only those sorts of work that coincided with a woman's natural sphere were to be encouraged. Such discrimination had little to do with the danger or unpleasantness of the work concerned. There was not much to choose for example – if our criterion is risk to life or health – between work in the mines, and work in the London dressmaking trades. But no one suggested that sweated needlework should be prohibited to women.

(Alexander, 1982, p. 33)

Thinking back to the contrast between the coalfields and the cotton towns and the relationship in each between economic structure and gender relations and roles, it is clear that the difference between the two areas was not simply based on the presence/absence of waged labour. We have, indeed, already suggested other elements, such as the whole ideology of virility attached to mining. But it was also to do with the *kind* of work for women in Lancashire: that it was factory work, with machines, and outside the home. In the sweated trades of nineteenth-century London, capitalism and patriarchy together produced less immediate threat to men's domination.

There were other ways, too, in which capitalism and patriarchy interrelated in the inner London of that time to produce a specific outcome. The sweated trades in which the women worked, and in particular clothing, were located in the inner areas of the metropolis for a whole variety of reasons, among them the classic one of quick access to fast-changing markets. But they also needed labour, and they needed cheap labour. Homeworking, besides being less of an affront to patriarchal relations, was one means by which costs were kept down. But costs (wages) were also kept down by the very availability of labour. In part this was a result of immigration and the vulnerable position of immigrants in the labour market. But it was also related to the predominantly low-paid and irregular nature of jobs for men (Harrison, 1983, p. 42). Women in Hackney *needed* to work for a wage. And this particular Hackney articulation of patriarchal influences and other 'location factors' worked well enough for the clothing industry.

But even given that in Hackney the social organization and nature of women's work was less threatening to men than in the cotton towns, there were still defensive battles to be fought. The labour-force of newly arrived immigrants also included men. Clearly, were the two sexes to do the same jobs, or be accorded the same status, or the same pay, this would be disruptive of male dominance. The story of the emergence of a sexual division of labour within the clothing industry was intimately bound up with the maintenance of dominance by males in the immigrant community. They did not use the confused and contradictory criteria of 'skill' and 'heavy work' employed so successfully in Lancashire. In clothing *any* differentiation would do. Phillips and Taylor (1980) have told the story, of the establishment of the sexual division of labour in production, based on the minutest of differences of job, changes in those differences over time, and the use of them in whatever form they took to establish the men's job as skilled and the women's as less so.

Rural life and labour

Our final example is drawn from the Fenlands of East Anglia, where the division of labour and gender relations took a different form again. In the rural villages and hamlets of nineteenth-century East Anglia, as in the Lancashire cotton towns, many women 'went out to work'. But here there was no coal industry, no factory production of textiles, no sweated labour in the rag trade. Economic life was still overwhelmingly dominated by agriculture. And in this part of the country farms were large, and the bulk of the population was landless, an agricultural proletariat. The black soils demanded lots of labour in dyking, ditching, claying, stone-picking and weeding to bring them under the 'New Husbandry', the nineteenth-century extension of arable land (Samuel, 1975, pp. 12 and 18). Women were an integral part of this agricultural workforce, doing heavy work of all sorts on the land, and provoking much the same moral outrage as did the employment of women in mills in Lancashire:

...the poor wage which most labourers could earn forced their wives to sell their labour too, and continue working in the fields. In Victorian eyes, this was anathema for it gave women an independence and freedom unbecoming to their sex. 'That which seems most to lower the moral or decent tone of the peasant girls', wrote Dr. Henry Hunter in his report to the Privy Council in 1864, 'is the sensation of independence of society which they acquire when they have remunerative labour in their hands, either in the fields or at home as straw-plaiters etc. All gregarious employment gives a slang character to the girls appearance and habits, while dependence on the man for support is the spring of modest and pleasing deportment'. The first report of the Commissioners on The Employment of Children, Young Persons and Women in Agriculture in 1867, put it more strongly, for not only did landwork 'almost unsex a woman', but it 'generates a further very pregnant social mischief by unfitting or indisposing her for a woman's proper duties at home'.

(Chamberlain, 1975, p. 17)

The social and spatial structure of the rural communities of this area also influenced the availability and the nature of work. Apart from work on the land, there were few opportunities for women to earn a wage. Even if they did not leave the village permanently, it was often necessary to travel long distances, frequently in groups, with even more serious repercussions in the eyes of the Victorian establishment.:

The worst form of girl labour, from the point of view of bourgeois respectability, was the 'gang' system, which provoked a special commission of inquiry, and a great deal of outraged commentary, in the 1860s. It was most firmly established in the Fen districts of East Anglia and in the East Midlands. The farms in these parts tended to be large but the labouring population was scattered...The labour to work the land then had to be brought from afar, often in the form of travelling gangs, who went from farm to farm to perform specific tasks.

(Kitteringham, 1975, p. 98)

There are here some familiar echoes from Lancashire. And yet things were different in the Fens. In spite of all the potential threats to morality, domesticity, femininity and general female subordination, 'going out to work' on the land for women in the Fens, even going off in gangs for spells away from the village, does not seem to have resulted in the kinds of social changes, and the real disruption to established ways, that occurred in Lancashire. In this area, women's waged-labour did not seem to present a threat to male supremacy within the home. Part of the explanation lies in the different nature of the work for women. This farm labour was often seasonal. The social and spatial organization of farmwork was quite different from that of factory work, and always insecure. Each gang negotiated wage rates independently with the large landowners, the women were not unionized, did not work in factories, were not an industrial proletariat in the same sense as the female mill workers in the cotton towns. Part of the explanation too, as in the colliery villages, lies in the organization of male work. Men, too, were predominantly agricultural labourers, though employed on an annual rather than a seasonal basis, and like mining, agricultural work was heavy and dirty, imposing a similar domestic burden on rural women.

A further influence was the life of the rural village, which was overwhelmingly conservative – socially, sexually and politically. Women on the land in this area did not become radicalized like women in the cotton towns. Relations between the sexes continued unchanged. Women served their menfolk, and both men and women served the local landowner; nobody rocked the boat politically:

When the Coatesworths ruled the village to vote Tory was to get and keep a job. The Liberals were the party of the unemployed and the undeserving...Concern over politics was not confined to men. The women took an interest, too. They had to. Their man's political choice crucially affected his employment, and their lives.

(Chamberlain, 1975, p. 130)

Where are they now?

What is life like in these areas now? Have the traditional attitudes about women's place in the home in the heavy industrial areas survived post-war changes? Have Lancashire women managed to retain the independence that so worried the Victorian middle class? In this century there have been enormous changes in many areas of economic and social life. The communications revolution has linked all parts of the country together, TV, radio, video and a national press have reduced regional isolation and increased the ease with which new ideas and attitudes spread. Changes in social mores, in the role of the family, in the labour process of domestic work, increased divorce rates and a rapid rise in women's participation in waged-labour between the Second World War and the end of the seventies have all had an impact. And yet, we shall argue here, regional differences remain.

There are, as we said in the introduction, two threads which we shall follow in this process of the reproduction of local uniqueness. The first concerns the geographically differentiated operation of national processes. Over 40% of the national paid labour-force in the UK now consists of women: a vast majority of them married. One of the consequences of this growth of jobs 'for women' has paradoxically been both an increase and a reduction in regional differences. The gender division of labour is changing in different ways in different areas, in part in response to previous patterns. Regional disparities in the proportion of women at work are closing, but the corollary of this, of course, is that the highest proportions of new and expanding jobs are in those very regions where previously few women have been involved in waged-labour. The four regions are being drawn in different ways into a new national structure of employment and unemployment. We cannot here attempt to explain this new spatial pattern. One thing we do hint at, though, is that the form of gender relations themselves, and the previous economic and social history of women in each of these places, may be one, though only one, thread in that explanation.

The areas, then, have experienced different types of change in their economic structure. In many ways the growth of jobs for women has been of greater significance in the north-east and in East Anglia than in the cotton towns or in Hackney. But that is not the end of the story. For those changes have themselves been combined with existing local conditions and this has influenced their operation and their effect. The impact of an increase in jobs for women has not been the same in the Fens as it has been in the coalfields of the north-east. This, then, is the second thread in our discussion of the reproduction of local uniqueness.

In the rest of this chapter we try to show the links between past and present patterns, how changing attitudes to women and men's roles at work and in the family in different parts of the country (themselves related to previous economic roles) both influence and are influenced by national changes in the nature and organization of paid employment over time. The present gender division of labour in particular places is the outcome of the combination over time of successive phases. Space and location still matter. The structure of relationships between men and women varies between, and within, regions. Life in inner London is still not the same as in the Fenlands, in the coalfields of the north-east, as in the textile towns round Manchester. The current division of labour between women and men is different, paid employment is differently structured and organized, and even its spatial form varies between one part of the country and another.

Coal was our life?

The decline of work in the pits is a well-known aspect of post-war economic changes in Britain. How have the men and women of the north-east reacted to this decline in their traditional livelihood? Have the changes challenged or strengthened the traditional machismo of the north-eastern male? What is happening in the north-east today in many ways recalls some of the images – and the social alarm – generated by the cotton towns a hundred years earlier. It is now in the north-east that homes are being 'turned upside down' and patriarchy threatened by women going out to work. At the beginning of the 1960s, still something less than a quarter of all adult women in the old colliery areas worked outside their homes for wages. The figure has more than doubled since then. And part of the explanation lies in the local distinctiveness, the uniqueness of these areas that has its origins in the nineteenth century. The women of this area have no tradition of waged-labour, no union experience. It was, of course, these very features that proved attractive to the female-employing industries that opened branch plants in increasing numbers in Co. Durham in the sixties and seventies.

The new jobs that came to the north-east, then, were mainly for women. They were located on trading estates and in the region's two New Towns built to attract industrial investment and also to improve housing conditions. The women who moved into the New Towns of Peterlee and Washington provided a cheap, flexible, untrained and trapped pool of labour for incoming firms. And added to this, the loss of jobs for men together with the rent rises entailed by a move to new housing pushed women into the labour market.

Male antagonism to the new gender division of labour was almost universal. Outrage at women 'taking men's jobs', pleas for 'proper jobs', an assumption that the packing, processing and assembly line work that loomed ever larger in the economic structure of the area was an affront to masculine dignity: "I think a lot of men feel that assembly work wouldn't be acceptable;

they'd be a bit proud about doing that type of work in this area. North East ideas are ingrained in the men in this area" (Lewis, 1983, p. 19). These assumptions appear to be shared by the new employers: "we are predominantly female labour orientated...the work is more suited to women, it's very boring, I suppose we're old-fashioned and still consider it as women's work...the men aren't interested".

This lack of interest plays right into the hands of the employers: once defined as 'women's work', the jobs are then classified as semi- or unskilled and hence low paid. An advantage that can be further exploited, as this factory director explains:

"we changed from full-time to part-time women(!)...especially on the packing... because two part-timers are cheaper than one full-timer...we don't have to pay national insurance if they earn less than £27.00 a week, and the women don't have to pay the stamp...the hours we offer suit their social lifes".

(Lewis, Ph.D., forthcoming)

So if men aren't doing jobs outside the house, what are they doing instead? Are men here, like their Lancashire forebears 'condemned to domestic occupations?'. Unlikely. An ex-miner's wife speaking on *Woman's Hour* in 1983 recalled that her husband would only reluctantly help in the home, pegging out the washing, for example, under cover of darkness!

Things *are* changing, though. Men are seen pushing prams in Peterlee, Newcastle-upon-Tyne Council has a women's committee, TV crews come to inquire into the progress of the domestication of the unemployed north-eastern male and the social and psychological problems it is presumed to bring with it. Working-class culture is still dominated by the club and the pub but even their male exclusivity is now threatened. The 1984 miners' strike seems set to transform gender relations evern further. New battle lines between the sexes are being drawn. The old traditional pattern of relations between the sexes, which was an important condition for the new gender division being forged in the labour market, is now under attack.

Industry in the country?

How has life changed in the Fens? In some ways, continuity rather than change is the link between the past and present here. For many women, especially the older ones, work on the land is still their main source of employment:

hard work, in uncompromising weather, in rough old working clothes padded out with newspaper against the wind...Marriage for convenience or marriage to conform... Land-worker, home servicer. Poverty and exploitation – of men and women by the landowners, of women by their men.

(Chamberlain, 1975, p. 11)

Not much different from their grandmothers and great-grandmothers before them. Gangs are still a common feature and the nature of fieldwork has hardly

Women harvesting in nineteenth-century Norfolk (reproduced by kind permission of the Coleman and Rye Library of Local History, Norwich).

changed either. Flowers are weeded and picked by hand. Celery and beet are sown and picked manually too. And this type of work is considered 'women's work'. It is poorly paid, seasonal and backbreaking. Male fieldworkers, on the other hand, have the status of 'labourers', relative permanence and the benefits associated with full-time employment. And they are the ones who have machinery to assist them.

Life *has* changed though. Small towns and rural areas such as the Fens have been favoured locations for the new branch plants and decentralizing industries of the sixties and seventies. Labour is cheap here – particularly with so few alternatives available – and relatively unorganized. Especially for younger women, the influx of new jobs has opened up the range of employment opportunities. It provides a means, still, both of supplementing low male wages, and of meeting people – of getting out of the small world of the village.

The impact of such jobs on women's lives, though, even the possibility of taking them, has been structured by local conditions, including gender relations. This is still a very rural area. The new jobs are in the nearby town. So unless factories provide their own transport (which a number do), access is a major problem. Public transport is extremely limited, and becoming more so. There are buses – but only once a week to most places. Not all families have a car, and very few women have daily use of one, let alone own 'their own' car. For many women, a bicycle is the only means of getting about.

This in turn has wider effects. For those who do make the journey to a factory job the effective working day (including travel time) can be very long. The

A landworker at Gislea Fen, 1974 (photograph by Angela Phillips, and reproduced with her kind permission).

time for domestic labour is squeezed, the work process consequently intensified. Those who remain in the village become increasingly isolated. The industrial workers, be they husbands or women friends, are absent for long hours, and services – shops, doctors, libraries – gradually have been withdrawn from villages.

It seems that the expansion of industrial jobs 'for women' has had relatively little impact on social relations in the rural Fens. In part, this is to do with the local conditions into which the jobs were introduced: the impact back of local factors on national changes. The Fenland villages today are still Conservative – politically and socially. Divorce, left-wing politics, women's independence are very much the exception.

Old cultural forms, transmitted, have remained remarkably intact:

Although love potions and true-lovers' knots made of straw have disappeared, Lent and May weddings are still considered unlucky. The Churching of Women – an ancient post-natal cleansing ceremony – is still carried on, and pre-marital intercourse and the resulting pregnancy is as much a hangover from an older utilitarian ap-proach to marriage as a result of the permissive society. In a farming community sons are important and there would be little point in marrying an infertile woman.

(Chamberlain, 1975, p. 71)

Attitudes to domestic responsibilities also remain traditional:

No women go out to work while the children are small – tho' there isn't much work anyway, and no facilities for childcare. Few women allow their children to play in the streets, or let them be seen in less than immaculate dress. Many men come home to lunch and expect a hot meal waiting for them. (p. 71)

It takes more than the availability of a few jobs, it seems, substantially to alter the pattern of life for women in this area:

Although employment is no longer dependent on a correct political line, the village is still rigidly hierarchic in its attitudes, and follows the pattern of the constituency in voting solidly Conservative. And in a rigidly hierarchical society, when the masters are also the men, most women see little point in taking an interest in politics, or voting against the established order of their homes or the community as a whole...Most women must of necessity stick to the life they know. Their husbands are still the all-provider. The masters of their lives.

(Chamberlain, 1975, pp. 130–1)

Gender relations in East Anglia apparently have hardly been affected by the new jobs, let alone 'turned upside down'.

A regional problem for women?

The contrast with the cotton towns of Lancashire is striking. Here, where employment for women in the major industry had been declining for decades, was a major source of female labour, already skilled, already accustomed to factory work, plainly as dexterous as elsewhere. And yet the new industries of the sixties and seventies, seeking out female labour, did not come here, or not to the extent that they went to other places.

The reasons are complex, but they are bound up once again with the intricate relationship between capitalist and patriarchal structures. For one thing, here there was no regional policy assistance. There has, for much of this century, been massive decline in employment in the cotton industry in Lancashire. Declines comparable to those in coalmining, for instance, and in areas dominated by it. Yet the cotton towns were never awarded Development Area status. To the extent that associated areas were not designated on the basis of unemployment rates, the explanation lies at the level of taxes and benefits which define women as dependent. There is often less point in signing on. A loss of jobs does not necessarily show up, therefore, in a corresponding increase in regional unemployment. Development Areas, however, were *not* designated simply on the basis of unemployment rates. They were wider concepts, and wider regions, designated on the basis of a more general economic decline and need for regeneration. To that extent the non-designation of the cotton towns was due in part to a more general political blindness to questions of women's employment.

So the lack of regional policy incentives must have been, relatively, a deterrent to those industries scanning the country for new locations. But it cannot have been the whole explanation. New industries moved to other non-assisted areas – East Anglia, for instance. Many factors were in play, but one of them surely was that the women of the cotton towns were not, either individually or collectively in their history, 'green labour'. The long tradition of women working in factory jobs, and their relative financial independence, has continued. In spite of the decline of cotton textiles the region still has a high female activity rate. And with this there continued, in modified form, some of those other characteristics. Kate Purcell, doing research in the Stockport of the 1970s, found that:

It is clear that traditions of female employment and current rates of economic activity affect not only women's activity per se, but also their attitudes to, and experience of, employment. The married women I interviewed in Stockport, where female activity rates are 45 per cent and have always been high, define their work as normal and necessary, whereas those women interviewed in the course of a similar exercise in Hull, where the widespread employment of married women is more recent and male unemployment rates are higher, frequently made references to the fortuitous nature of their work.

(Purcell, 1979, p. 119)

As has so often been noted in the case of male workers, confidence and independence are not attributes likely to attract new investment. It may well be that here there is a case where the same reasoning has applied to women.

But whatever the precise structure of explanation, the women of the cotton towns are now facing very different changes from those being faced by the women of the coalfields. Here they are not gaining a new independence from men; to some extent in places it may even be decreasing. Women's unemployment is not seen to 'disrupt' family life, or cause TV programmes to be made about challenges to gender relations, for women do the domestic work anyway. Having lost one of their jobs, they carry on (unpaid) with the other.

Hackney: still putting out

What has happened in Hackney is an intensification of the old patterns of exploitation and subordination rather than the superimposition of new patterns. Here manufacturing jobs have declined, but the rag trade remains a major employer. The women of Hackney possess, apparently, some of the same advantages to capital as do those of the coalfields and the Fens: they are cheap and unorganized (less than 10% are in a union – Harrison, 1983, pp. 69–70). In Inner London, moreover, the spatial organization of the labour-force, the lack of separation of home and work, strengthens the advantages: overheads (light, heat, maintenance of machinery) are borne by the workers themselves; workers are not eligible for social security benefits; their spatial separation one from another makes it virtually impossible for them to combine to force up wage rates, and so on.

So given the clear advantages to capital of such a vulnerable potential workforce, why has there been no influx of branch plants of multinationals, of electronics assembly-lines and suchlike? Recent decades have of course seen the growth of new types of jobs for women, particularly in the service sector, if not within Hackney itself then within travelling distance (for some), in the centre of London. But, at the moment, for big manufacturing capital and for the clerical–mass production operations which in the sixties and seventies established themselves in the Development Areas and more rural regions of the country, this vulnerable labour of the capital city holds out few advantages. Even the larger clothing firms (with longer production runs, a factory labour process, locational flexibility and the capital to establish new plant) have set up their new branch plants elsewhere, either in the peripheral regions of Britain or in the Third World. So why not in Hackney? In part the women of Hackney have been left behind in the wake of the more general decentralization, the desertion by manufacturing industry of the conurbations of the First World. In part they are the victims of the changing international division of labour within the clothing industry itself. But in part, too, the reasons lie in the nature of the available labour. Homeworking does have advantages for capital, but this way of making female labour cheap is no use for electronics assembly-lines or for other kinds of less individualized production. The usefulness of this way of making labour vulnerable is confined to certain types of labour process.

The influx of service jobs in central London has outbid manufacturing for female labour, in terms both of wages and of conditions of work (see Massey, 1984, ch. 4). But working in service jobs has not been an option available to all. For women in one way or another tied to the home, or to the very local area, homeworking in industries such as clothing has become increasingly the only available option. Given the sexual division of labour in the home, homeworking benefits some women:

Homework when properly paid, suits many women: women who wish to stay at home with small children, women who dislike the discipline and timekeeping of factory work and wish to work at their own pace. Muslim women observing semi-purdah.

(Harrison, 1983, p. 64)

But homework seldom is 'properly paid'. Harrison again, on types of work and rates of pay in Hackney in 1982:

There are many other types of homework in Hackney: making handbags, stringing buttons on cards, wrapping greeting cards, filling Christmas crackers, assembling plugs and ballpens, sticking insoles in shoes, threading necklaces. Rates of pay vary enormously according to the type of work and the speed of the worker, but it is rare to find any that better the average female hourly earnings in the clothing trade in 1981, £1.75 an hour, itself the lowest for any branch of industry. And many work out worse than the Wages Council minimum for the clothing trade of £1.42 per hour (in 1982). Given these rates of pay, sometimes the whole family, kids and all, are dragooned in:...one mother had her three daughters and son helping to stick eyes and tails on cuddly toys.

(Harrison, 1983, pp. 67–8)

The involvement of all members of a family in homework or working as a team in small family-owned factories is not uncommon, especially among ethnic minorities. For small companies the extended family may be essential to survival:

the flexibility comes from the family: none of their wages are fixed. When times are good, they may be paid more. When they are bad, they are paid less. They get the same pay whether their hours are short or long.

The fact that women are employed in the context of an extended family is important not only in the organization of the industry but also for the lives of the women themselves. They may have a wage, but they do not get the other forms of independence which can come with a job. They do not get out of the sphere of the family, they do not make independent circles of friends and contacts, nor establish a spatially separate sphere of existence. Within the family itself the double subordination of women is fixed through the mixing in one person of the role of husband or father with that of boss and employer.

But it is not that there have been no changes in recent decades for the homeworkers of Hackney. They too have been caught up in and affected by the recent changes in the international division of labour. The clothing industry of London in the second half of the twentieth century finds itself caught between cheap imports on the one hand and competition for labour from the better working conditions of the service sector on the other. The clothing firms with the ability to do so have long since left. For those that remain, cutting labour costs is a priority, and homeworking a means to do it. So an increasing proportion of the industry's work in the metropolis is now done on this social system while the amount of work overall, and the real wages paid, decline dramatically. For the women who work in this industry there is thus more competition for available work, increasing vulnerability to employers and intensification of the labour process. And this change in employment conditions brings increased pressures on home life too, though very different ones from those in the north-east, or the Fens. For these women in Hackney their workplace is also their home.

Here's Mary, a forty-five-year-old English woman with teenage children describing the pressures she feels:

I've been machining since I was fifteen, and with thirty years' experience I'm really fast now...But I'm having to work twice as hard to earn the money. The governors used to go on their knees to get you to take work if they had a rush to meet a delivery date. But they're not begging no more. It's take it or leave it. If you argue about the price they say we can always find others to do it. It's like one big blackmail. Three years ago we used to get 35p to 40p for a blouse, but now [1982] you only get 15p to 20p...

I used to get my work done in five hours, now I work ten or twelve hours a day...The kids say, mum, I don't know why you sit there all those hours. I tell them, I don't do it for love, I've got to feed and clothe us. I won't work Sundays though. I have to think about the noise...I'm cooped up in a cupboard all day – I keep my machine in the storage cupboard, it's about three feet square with no windows. I get pains in my shoulders where the tension builds up. I've got one lot of skirts to do now, I've got to do sixteen in an hour to earn £1.75 an hour, that means I can't let up for half a second between each skirt. I can't afford the time to make a cup of tea. With that much pressure, at the end of the day you're at screaming pitch. If I wasn't on tranquillizers, I couldn't cope. I'm not good company, I lose my temper easily. Once I might have been able to tolerate my kids' adolescence, with this I haven't been able to, I haven't been able to help them – I need someone to help me at the end of the day.

(Harrison, 1983, pp. 65–7)

Reflected in this woman's personal experience, her sweated labour and family tensions, is a new spatial division of labour at an international scale. Low wage, non-unionized workers in Hackney are competing directly with the same type of low-technology, labour-intensive industries in the Third World. But it is precisely the history of the rag trade in Hackney, the previous layers of economic and social life, that have forced this competition on them. The intersection of national and international trends, of family and economic relationships, of patriarchy and capitalism have produced this particular set of relationships in one area of Inner London.

References

Alexander, S. (1982) 'Women's work in nineteenth-century London: a study of the years 1820–50', pp. 30–40 in E. Whitelegg *et al.* (eds.), *The Changing Experience of Women*, Martin Robertson, Oxford.

Anderson, M. (1971) *Family and Structure in Nineteenth-Century Lancashire*, Cambridge University Press, Cambridge.

Chamberlain, M. (1975) *Fenwomen*, Virago, London.

Engels, F. (1969 edn) *The Conditions of the Working Class in England*, Panther, St Albans.

Frankenberg, R. (1976) 'In the production of their lives, man (?)...sex and gender in British community studies', chapter 2, pp. 25–51 in D. L. Barker and A. Allen (eds.), *Sexual Divisions and Society: Process and Change*, Tavistock, London.

Hall, C. (1982) 'The home turned upside down? The working class family in cotton textiles 1780–1850', in E. Whitelegg *et al.* (eds.), *The Changing Experience of Women*, Martin Robertson, Oxford.

Harrison, P. (1983) *Inside the Inner City*, Penguin, Harmondsworth.

Kitteringham, J. (1975) 'Country work girls in nineteenth-century England', Part 3, pp. 73–138, in R. Samuel (ed.), *Village Life and Labour*, Routledge and Kegan Paul, London.

Lewis, J. (1983) 'Women, work and regional development', *Northern Economic Review*, no. 7. Summer, pp. 10–24.

Lewis, J. (forthcoming) Ph.D Thesis, Department of Geography, Queen Mary College, London.

Liddington, J. (1979) 'Women cotton workers and the suffrage campaign: the radical suffragists in Lancashire, 1893–1914', chapter 4, pp. 64–97, in S. Burman (ed.), *Fit Work for Women*, Croom Helm, London.

Massey, D. (1984) *Spatial Divisions of Labour: Social Structures and the Geography of Production*, Macmillan, London.

Phillips, A. and Taylor, B. (1980) 'Notes towards a feminist economics', *Feminist Review*, vol. 6, pp. 79–88.

Purcell, K. (1979) 'Militancy and acquiescence amongst women workers', chapter 5, pp. 98–111, in S. Burman (ed.), *Fit Work for Women*, Croom Helm, London.

Samuel, R. (1975) *Village Life and Labour*, Routledge and Kegan Paul, London.

Strong Words Collective (1977) *Hello, are you working?* Erdesdun Publications, Whitley Bay.

Strong Words Collective (1979) *But the world goes on the same*, Erdesdun Publications, Whitley Bay.

Webb, S. (1921) *The Story of the Durham Miners*, Fabian Society, London.

8

The laissez-faire approach to international labor migration: the case of the Arab Middle East*

ALAN RICHARDS AND PHILIP L. MARTIN

An estimated 14–20 million persons are currently living and working in countries where they are neither citizens nor immigrants. Half of these nonimmigrant workers are legally admitted 'guestworkers'; the rest are 'illegal aliens' or 'undocumented workers.' These migrant workers must be distinguished from two other transient groups: the 1 million permanent immigrants who begin anew in another country each year and the 13 million refugees living outside their country of citizenship and liable to prosecution if they return. The distinctions between the three groups are often blurred, as when migrant workers become immigrants.

The migration familiar to Americans moved transients and settlers from East to West. The migratory chain established in the nineteenth century recurs today – single males migrate first and later are joined by their dependents. Family reunification and formation establish a community in the receiving area to which later migrants come. Thus is forged the migratory chain which moves people between two areas. From 1800 to 1920, some 50 million Europeans arrived in the Americas. Early waves of immigrants intended (or were forced) to effect a relatively clean break with their homeland.

Migration streams mature over time. The second wave of immigrants in the late nineteenth and early twentieth centuries contained many 'target earners'; young men who hoped to work hard, live frugally, save money, and return home to marry, buy a farm, build a house, or open a small store.[1] Of course, many never returned. Most of today's migrants are also target earners – skilled and unskilled laborers moving from poor to rich countries. Since there are relatively few permanent immigrant slots, most workers have

* Source: *Economic Development and Cultural Change*, vol. 31, no. 3, April 1983, pp. 455–71. © University of Chicago Press, 1983. All rights reserved.

Editors' note: we have omitted the majority of the references and detailed footnotes in the original. Readers interested in the sources used by the authors should refer to the journal.

no choice but to be temporary workers, often moving back and forth between their home country and their place of work.

Economists believe that voluntary migration benefits not only individual migrants and employers but also sending and receiving countries. Drawing on the theory of competitive equilibrium, usually in the form of simple international trade theory, they commonly assert that since labor is a commodity like any other, if two nations have unequal resource endowments exchange is mutually beneficial. The importing country is able to fill job slots at a lower cost than would otherwise be possible, which reduces inflationary pressures. Labor-importing countries are thought to derive dynamic benefits as well: flexible and elastic labor supplies allegedly prevent industrial expansion from bidding up wages, reducing profits, and retarding investment.[2]

Exporting countries are also believed to gain: by exporting a relatively abundant factor (labor), they raise home wages and generate a return flow of human and financial capital. Migration tends to equalize input and output prices, increasing efficiency and welfare for all concerned. In this view, economic benefits are maximized by minimizing the barriers to migration, a laissez-faire policy. Free trade in labor is no different from free trade in goods, and both are desirable.

For years some countries have followed the economists' lead in endorsing and encouraging international labor migration. Industrial nations thought they could obtain the additional labor needed to sustain noninflationary growth, and labor-exporting nations hoped to reduce unemployment and obtain remittance incomes. Recently both sending and receiving countries have reversed their previous policies: laissez-faire has fewer supporters these days. Labor importers found that migrants did not solve basic structural problems. Instead, the presence and availability of migrants may preserve low-wage, labor-intensive industries and make it more difficult to reduce trade barriers or promote productivity-increasing innovations. Many people began to feel that it was "morally wrong to build the development of our wealth on the backs of foreign manpower...a group of people who are identifiably of another race to do the despised menial work."[3]

Sending countries also began to question the wisdom of laissez-faire policies which sent the "best and brightest" abroad more or less permanently. Algeria and Yugoslavia, for example, have drastically reduced labor emigration. The government of South Yemen has prohibited labor emigration altogether. Even countries with a free enterprise ideology, such as the Kingdom of Jordan, have called for an international fund to compensate sending countries for the losses that labor exports impose upon them.[4]

What went wrong with the laissez-faire policy prescription? Why were the expectations created by orthodox theory not fulfilled? This paper examines these questions for sending countries by reviewing contemporary labor migration in the Middle East. This region provides a useful case study for an analysis of laissez-faire policies. First, labor flows are quite large: at least 3 million aliens are living and working in the principal receiving countries. Second, although receiving countries have placed some legal restrictions on labor migration, these are often unenforced, while the sending countries of the region have until recently pursued almost textbook laissez-faire policies. An analysis of the Middle Eastern case not only provides insights into the development dilemmas in this vital region but also may help to pinpoint the weaknesses of laissez-faire theories and policies on labor migration.

Middle East labor migration: an overview

Estimates of Middle East labor flows vary considerably. The most comprehensive survey to date is that of Birks and Sinclair.[5] Their numbers should probably be regarded as lower bounds, even for their 1975 cut-off point. The strength of their estimates lies in the fact that they cross-check the claims of sending and receiving countries. However, it is widely believed that their estimates are too low. For example, the major exporters of skilled labor are Egypt and Yemen. Choucri, Eckaus, and Mohi el-Din believe that at least 1 million Egyptians were abroad in 1978 (as opposed to Birks and Sinclair's estimate of 400,000), while the Egyptian government places the figure at 1.2 million. The World Bank estimates the numbers of North Yemenis abroad at over 1.2 million in 1978. Higher estimates for the main labor importers place Saudi Arabia's migrant population at 1.5 million, Libya's at 0.5 million, the United Arab Emirates' at 400,000, and Kuwait's at 350,000, with smaller numbers in Qatar and Bahrain. Some countries, like Algeria, Iraq, and especially Jordan and Oman, both import and export labor. Because census avoidance is widespread in the Middle East, and because the situation is changing rapidly, these numbers can only give us a very general notion of the magnitude of the flows.

Receiving countries are highly dependent on migrants, who often comprise more than 50% of the work force. Figure 8.1. provides comparisons of migrant workers to domestic populations, illustrating the fact that migrant work forces exceed the domestic populations of the UAE, Qatar, and Kuwait, while the migrant work forces of Saudi Arabia, Bahrain, and Libya are 30%–40% of the host country's total population. Estimates of the share and distribution of migrant workers are only approximations, but it appears that the UAE, with a work force that is 90% foreign, has the highest migrant dependence ratio. In Kuwait, where oil was discovered in 1946 and where 50% of the population was foreign in 1958, the work force today is about 80% foreign, the same percentage as in Saudi Arabia.

Within the Middle East, migration for employment occurs in a fundamentally laissez-faire environment. Although there are stringent restrictions on migration for settlement, labor emigration is relatively unimpeded. Egypt places no formal barriers in a migrant's path; Yemenis do not need work

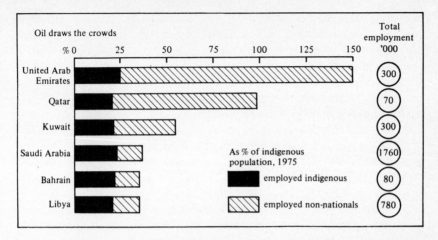

Fig. 8.1. Comparisons of migrant workers to domestic populations in various oil-producing countries. (Source: *Third World Quarterly*, April 1979)

permits in Saudi Arabia. Those restrictions which do exist often go unenforced; for example, Egyptian migration to Libya, occasionally 'prohibited' for political reasons, continues either clandestinely or by migration first to a third country (typically Tunisia) and then on to Libya. The Yemen Arab Republic's attempt to limit migration in order to increase the pool of men of military age is proving impossible to enforce because the central government has little or no control over the northern tribal areas, which border Saudi Arabia.

Two forces have produced this environment: (1) a migratory tradition and (2) the fundamental transformations which the oil boom has wrought in the political economy of the region. As many historians of the Islamic world have pointed out, educated Muslims have long travelled freely from one end of the *Dar al-Islam* (House of Islam) to the other. Speaking of medieval Islam, Marshall Hodgson wrote: "Practically every well-known Muslim lived in many cities: soldiers travelled...in the way of conquest; scholars travelled to find new teachers and new libraries and also to find more appreciative audiences."[6] Nor was such movement limited to elites: the *hajj*, or pilgrimage, is one of the five Pillars of Islam; millions of Muslims from virtually all social strata have completed the ritual. It is almost certainly the largest multinational gathering on earth, with some 1.5 million pilgrims coming every year.

National boundaries and the nation-state itself coexist uneasily with the tradition of mobility, rooted in long-distance commerce, and with Islamic political thought. Although in fact different political units have characterized the Muslim world almost from its inception, such realities have been rationalized as necessary evils. Cosmopolitan mobility, not civic loyalty, has been accorded primary legitimacy. The parallel yet distinct modern heritage of pan-Arab nationalism further weakens the legitimacy of restrictions on labor migration. This highly influential ideology holds that all boundaries from Morocco to the Shatt-al-Arab are artificial: there is only one indivisible Arab nation. It is true that the Arab region is bound by common ties of language, religion, and social custom. Yet, as with Islamic political theory, there has long been a gap between theory and reality. Arab nation-states obviously exist, and attempts at unity have repeatedly foundered. Ironically, the most passionately Arab nationalist regimes, such as Nasser's Egypt or Baathist Syria, have placed the most restrictions on the movement of labor within the Arab world. Nevertheless, the ideological view of artificial borders, coupled with Islamic beliefs and practices, provides considerably less legitimacy for restricting labor migration than do the political traditions of, say, Western Europe.

But fundamentally it has been economic forces, not religion or ideology, that have shaped the size and structure of contemporary labor migration. The 10-fold increase in oil prices since 1973, ambitious development plans, and small populations provided the motor for an accumulation process that required external labor. Although large sums were devoted to importing military hardware and Western consumption (often luxury) goods, all oil exporting countries have established and at least partially implemented extensive economic development programs. Saudi Arabia spent $180 billion from 1975 to 1980 and plans to spend some $290 billion more by 1985. Libya spent $39.9 billion from 1973 to 1980. Even small states like Qatar and the UAE devoted considerable sums – $10 billion and $9 billion, respectively – to development. This created a very strong demand for labor in the oil exporting countries.

Small populations, the low rate of labor force participation by women, and the aversion of many Bedouins to manual labor meant that the supply of labor from the indigenous population fell far short of this demand. The oil boom itself helped to create other sources of supply: by shifting the political balance of forces in the region away from radical nationalist regimes and toward conservative and traditionalist ones, the oil boom contributed to the shift in the foreign economic policy of Egypt, the number one labor exporter in the area. The process began in 1967 but greatly accelerated after 1973, culminating in Sadat's market-oriented *Infitah* (opening up) policies. Most commentators have stressed the resulting inflow of foreign goods; perhaps more important has been the outflow of people. The open door swings both ways.

The oil boom reinforced and extended the previous pattern of large-scale emigration of skilled and professional workers but also created largely new flows of semiskilled and unskilled workers. As noted above, there is a very long tradition of educated labor migration in the region. In the post-World War II era, such flows have been primarily movements of skilled, professional Egyptians, Palestinians, and Lebanese into the oil-rich states. This is under-standable, given educational traditions in the sending countries. The Palestinians, of course, had little choice but to seek employment abroad.

Remunerative jobs for the educated were also scarce in Lebanon and especially in Egypt.

The oil boom reinforced this trend. The oil states desperately need high-level, Arabic-speaking workers. All of the OAPEC (Organization of Arab Petroleum Exporting Countries) nations have embarked on large-scale expansions of their educational systems; most of the teachers are Egyptians or Levantines. Further, since oil wealth flows directly into the coffers of the state and since all OAPEC states need highly skilled technocrats to supervise their vastly expanded development plans and projects, Arab migration for government employment has likewise increased. The high salaries available in the oil countries, coupled with an expansion in their demand for high-level workers, strengthened and augmented a well-established pattern of migration.

The oil boom also stimulated large flows of less skilled labor. Such workers range from building craftsmen to common laborers. They are employed primarily in services and construction. Indeed, construction workers form a significant proportion of the total work force. As many as one-third of the 300,000 Egyptians estimated to be in Saudi Arabia are employed in construction. Almost 29% of the nonagricultural Saudi work force was employed in construction in 1975 (5% in the United States). Most of these construction workers are migrants, usually Egyptians, Yemenis, and, increasingly, non-Arab Asians. As we shall see in the next section, the high proportion of construction workers in the total flow of migrants has very important implications for future migrant labor needs. Many, perhaps most, of the migrants are building factories and infrastructure that will require few workers to operate and maintain. Western Europe, in contrast, imported migrant labor to staff labor-intensive factories and services on a continuous basis.

The labor flows in the Arab Middle East differ from those in Western Europe or the United States in several other respects as well. Unskilled migrants in all of these cases typically fill jobs that local workers disdain. But in the advanced industrial countries, prolonged economic growth and structural change have generated complex job hierarchies and stimulated a desire among workers for upward mobility. Native workers, who are often the children or grandchildren of migrants, want jobs with higher status and pay, leaving openings at the bottom of the hierarchy to be filled by new migrants.

No such historical process has occurred in the Middle East. There the native workers' disdain for manual labor derives from preindustrial social norms and from the role of a paternalistic state. The age-old symbiotic tension between agriculturalists and pastoralists in the region underlies the latter's rejection of manual labor. Former Bedouins typically become soldiers or drivers, shunning manual labor as a task for *fellahin* (peasants) and thus beneath their dignity. Consequently, a principal potential source of manual labor has bypassed *any* industrial work rather than having 'moved through' it. Direct government payments and subsidies for housing, medical care, education, and other services further reduce the incentives for the local population to assume jobs in the construction or service sectors.

A final distinctive feature of Mideast labor migration should be noted. Unlike flows from the Mediterranean to Northern Europe or from Latin America to the United States, Mideast workers are not moving from structurally less developed countries to more developed areas. Indeed, for Egyptians, Palestinians, and Lebanese, the reverse is true: workers move from their homelands of higher literacy and more developed industry (especially true for Egypt) to less industrialized, less well-educated nations. Workers move to oil rich – not highly industrialized – economies. This implies, of course, that one of the commonly alleged benefits of labor migration, acquisition of skills, has little relevance for the Middle Eastern case.

In summary, labor flows in the Middle East (1) bulk very large in the labor markets of the importing countries, (2) occur in a basically laissez-faire environment, (3) comprise both highly skilled and unskilled labor, (4) contain a relatively high proportion of workers producing investment (largely construction) goods, (5) fill jobs which locals either are untrained for or disdain because of preindustrial tradition and state policy, and (6) move from poor to rich countries but *not* from structurally less developed to more industrialized nations. We now turn to the problems such labor flows have created for the sending countries.

Problems of labor migration for sending countries

Labor migration has created three types of problems for the principal sending countries: (1) uncertainty for both economic and political reasons about the continuation or expansion of the current flows; (2) micro effects of remittance spending and labor migration; and (3) the selectivity of migration and its impact on certain key sectors. We examine these in turn.

A principal difficulty facing planners in a labor sending country is uncertainty about how long the main benefits of migration (namely, reduced unemployment and inflows of remittances) will last. Regardless of whether or not the remitted funds are in a form directly usable for investment spending (an issue examined below), planners need to have a fairly clear idea how much money will be flowing out and how much flowing in if they are to formulate realistic development plans and projects. Unfortunately, these flows in the Middle East are subject to political and economic uncertainty. In all cases labor migration is a response to increased demand for specific kinds of labor. Labor flows may be affected both by changes in the aggregate level of economic activity and by shifts in the composition of the demand for labor. [...]

It is probable that the rate of growth of oil government revenues will not increase at the same speed during the 1980s as in the middle and late 1970s. This alone will lead to a reduction in demand for migrant labor, unless we assume that the composition of demand will shift toward more labor-intensive techniques and commodities. However, precisely the opposite shift seems more likely.[...]

In the Middle East, there is a further issue: much of the current investment is in construction. Because of construction's very long life, at some point, even abstracting from financially generated 'busts' in construction typical of more advanced market economics, the demand for construction labor must decline. Further, it is clear that oil exporting countries are building energy- and petroleum-intensive industries, such as oil refining, ammonium fertilizer, and aluminium refining and fabrication. Such plants use very little labor. This pattern of investment leaves little room for continued, much less expanded, labor migration. Some observers have also observed a tendency for construction techniques to become increasingly capital intensive.

Two caveats are in order. While it seems clear that the demand for construction labor will decline at some point, it is unclear just when this will occur. This, of course, is part of the problem: the uncertainty of the foreign demand for labor upon which several exporting countries have come to depend. The evidence on the length of the construction boom is mixed. Some observers predicted that Saudi construction spending will actually decline some 15% over the next 2 years, but the recently unveiled Five Year Plan projects very large increases in construction. Nevertheless, there can be little doubt that construction spending will slow down, even in Saudi Arabia, during this decade. Recent research by over 150 British banks with Middle Eastern branches predicts a deceleration in construction activity in the principal labor-importing states, simply because basic infrastructure is now well in place. Kuwait, a more structurally developed oil exporter, shows the others what their future may be. Although construction occurs in Kuwait, the rate of increase in current government expenditure is merely keeping up with inflation and relatively few new investment projects are planned. The economy seems to be settling into the role of 'mature rentier.'

Further, even if there were no economic reasons to expect a decline in the demand for imported labor, there are political reasons to anticipate moves in that direction. The 'Iranian model,' of course, stands out as an example of a disruptive transformation. The recent conspiracy against the House of Saud that culminated in the occupation of the Great Mosque in Mecca underlines the dangers of rapid structural transformation which offends local mores and leaves a substantial portion of the rural population behind. Especially younger members of the elite, whose influence can only increase with time, do not want to inherit oil fields pumped dry, bank accounts ravaged by inflation, industrial facilities not competitive in world markets, and societies so churned up that their own positions would be much eroded.

Host country governments clearly perceive migrants as a necessary evil.

Arab migrants in particular are viewed with suspicion. The case of the Palestinians is the most obvious: nervousness over their role in Kuwait is endemic in ruling circles. Egyptian migrants face the delicate problem that their main destination, Libya, is now perceived as the principal enemy by the Egyptian military, while next-in-line Saudi Arabia is at odds with Cairo over the Camp David treaty. Yemeni–Saudi antagonism dates at least to the latter's seizure of Asim province in the 1920s. It is reinforced by the disdain with which many Saudis treat Yemeni manual workers. It may be said that antagonism toward the Saudis is one of the ties binding the otherwise ideologically hostile regimes of North and South Yemen. The Saudis, in turn regard the Yemenis as potential subversives, since Yemeni migrants often have strong republican sympathies.

The Saudi response has been an increasing tendency to import non-Arab labor, such as Thais, Filipinos, and Koreans. The latter are especially favored. In 1979, for example, Korean firms won *all* the new construction contracts let in Saudi Arabia. Korean firms now have nearly one-quarter of the total Middle Eastern construction market. Because Korean firms provide most of their own laborers, who work very long hours and live in isolation from the local population, their increasing popularity in the politically jittery Kingdom is not surprising. Should this trend continue, it would ensure a slowdown in the rate of in-migration of Arab labor. Even a net decline in the number of Arab workers in the oil countries cannot be ruled out. So far economic need has restrained politically motivated expulsion of migrant labor but it is a possibility that workforce planners in this volatile region cannot ignore. Political uncertainty reinforces economic uncertainty; both raise serious doubts about the long-run viability of large-scale Arab migration for employment.

Such uncertainty also surrounds the return flow of remittances. So long as such flows continue, their macroeconomic impact seems clearly beneficial. This effect is independent of the micro effects of remittances and the extent to which governments can tap these funds directly. By relaxing the foreign exchange constraint, such flows improve receiving countries' international credit positions; governments have an expanded capacity to borrow for development projects or alternatively, can reduce their foreign indebtedness. Governments are then free to concentrate their energies and funds on economic development projects rather than worrying constantly about the next debt payment. This seems to have occurred in Egypt: foreign debt fell from $4 billion to $2 billion between 1975 and 1979, largely as a result of worker remittances. Of course, these benefits accrue only so long as workers remain abroad.

The uncertainty surrounding these flows reduces their usefulness for development planning. If a regime incurs debts on the basis of such flows, it may create serious problems for the future if its expectations are not fulfilled. For example, Turkey embarked upon ambitious development plans while 650,000 Turkish workers were abroad, borrowing from foreign banks in the process. Turkish foreign debt now exceeds $14 billion (half of Turkey's export earnings) just as the return flow of remittances has been reduced. A regime which depends largely on workers' remittances as a source of foreign exchange is in the same position as any other one-commodity exporter. Unstable, fluctuating remittances are no more an unmixed blessing than unstable, fluctuating sugar sales.

We now turn to the second set of problems which surround labor migration: the microeconomic impacts of labor and remittance flows. [...] Although there is little direct evidence for the Middle East, the Mediterranean and Mexican experiences indicate that remittance funds generally flow into consumption rather than investment. Much of the investment that does occur is in housing. There are several reasons why such spending is rational from the point of view of the individual migrant. First, many of the migrants are very poor and quite naturally tend to spend foreign earnings to increase their immediate standard of living. Second, even if they should have a preference for saving, financial institutions in their home countries are typically very weak, especially in the rural areas; there is often no efficient vehicle for saving. Third, many of the needed investments in the rural (and some urban) areas are collective goods – wells, sewage systems, irrigation networks, roads, etc. Since remittances typically flow into rural home communities in small amounts, and it is quite rational for migrants or their families to spend money on personal consumption goods, there is little evidence that migrant remittances are available for the kind of investment spending which many of these areas need.

Personal consumption spending should not automatically be condemned – individual migrants and their families are clearly better off. It is also possible that such spending has beneficial social effects. The size of the income and employment multipliers depends on the import content and the labor intensity of locally produced goods. What little evidence there is suggests that Middle Eastern migrants, like those in other parts of the world, spend their incomes on improved food, clothing, housing, and household effects. The economic impact of such spending varies from country to country, but in general only the last two kinds of spending seem to be of the labor-intensive employment-generating type.

In some countries (e.g., Yemen) emigration is so massive that local labor cannot easily provide newly demanded goods, prompting increased imports. This is not true for housing, a nontradable, but appears to be so for the other major commodity categories, especially for food. Remittances lead to an increase in the demand for high value crops, such as vegetables. But since these are labor intensive, and since labor is often not available or is very expensive because of emigration, the increased demand is supplied by imports. In Yemen the value of food imports has increased 10-fold from 1971–2 to 1978. In Egypt also, increased remittances have stimulated food imports. The economic and political risks of such increased reliance on (largely Western) food imports for Middle Eastern countries are vividly illustrated by the suspension of US food exports to the USSR and the frequent proposals to cut them off to Iran.

This agricultural impact is the result, not merely of increased demand, but also of bottlenecks in supply. Such effects seem to induce agricultural mechanization in Egypt, Yemen, and Oman. The increase in tractor use due to wage increases is not surprising. The income effect of increased farm family income from remittances may also contribute to mechanization.

At first glance, tractors appear to be an important advance for agricultural development and are so treated by numerous authors. Yet there are several potential problems with agricultural mechanization in Middle Eastern countries. First, there are serious maintenance problems, since those with mechanical skills are most likely to migrate. Second, agricultural machinery, especially tractors, typically has a high import content. Third, and most important, there is the uncertainty problem again. Agricultural mechanization, like many technical changes, is usually an irreversible phenomenon. If a million migrants returned, countries like Egypt would be stuck with technologies highly inappropriate for their changed factor endowments. Some kinds of mechanization may actually undermine long-run agricultural production potential through misuse. In both Oman and Iran, the purchase of internal-combustion water pumps has led to overexploitation of ground water and the decline and collapse of older irrigation systems. Labor migration itself may have other detrimental consequences for food production, as terraces fall into disrepair (Yemen), irrigation systems are not maintained (Oman and Iran), and farmers shift to labor-saving, nonfood crops (*qat* in Yemen). Mechanization may maintain food production by factor substitution, but its relative irreversibility may cause future problems.

The selectivity of labor emigration may exacerbate these problems. The skill and age composition of migrants, coupled with low substitutability among different categories of workers, can cause serious supply bottlenecks in sending countries. The magnitude of this effect will depend on the speed with which new workers can acquire the necessary skills.[...]Migration selectivity may mean the loss of highly skilled workers. The 'brain drain,' of course, has long been a concern of sending countries throughout the world. One might argue that such problems should not arise in a country such as Egypt, which bursts at the seams with educated, underemployed labor. No doubt there are indeed some benefits in exporting such workers. It is still likely, however, that the best professionals depart; their special talents and skills are then lost to the home country. Insofar as this occurs in Egypt, the home government's policies may help to push out such talent. Not only are wages very low relative to OPEC countries, but the strict seniority pay system provides few incentives or challenges to the most productive professionals.

Conclusion

The Middle East provides an interesting test cast of laissez-faire migration policies and theories. We have argued that, while individual migrants and individual employers obviously benefit, the impacts on sending societies as a whole are not so unambiguously benign. Such a disjuncture between individual and social welfare presumably would not occur in an environment where all of the conditions for a competitive equilibrium are present. If we are correct that social and individual costs and benefits diverge, then we must be able to point to departures from the assumptions of competitive equilibrium theory in the realities of Middle Eastern labor migration.

We find five such divergences: (1) widespread uncertainty; (2) less rapid growth of demand for labor in receiving countries as infrastructure is put in place; (3) the nature of labor power, that is, the fact that workers cannot be separated from their work and can possess destabilizing political convictions; (4) problems of investment opportunities, factor proportions, and the like (usually due to a market imperfection) that reduce the volume and distort the structure of job-creating investment financed by remittances; (5) the nature of labor markets and technical irreversibilities in agriculture leading to patterns of supply and demand in the agricultural sector that may not be viable over the long term. Any of these features taken singly would be sufficient to weaken severely the relevance of models in which an unaided price system generated an optimal outcome. Taken together, the five divergences make laissez-faire policies, necessarily based upon such a model, highly questionable.

None of this means that labor flows should be deliberately reduced or stopped by sending governments (although some, like Yemen and Algeria, have taken steps in this direction). Nor do we propose an alternative policy at the same level of generality or alleged universal applicability as laissez-faire. Rather, we are arguing that such a general theoretical framework is unhelpful. The problems of labor migration arise from the specificity of the political and economic problems of both sending and receiving countries. The appropriate policies should be equally specific. They would, however, be policies, not the absence of policy implied by the laissez-faire model.

Notes

* This research was partially supported by a grant from the Agricultural Development Systems Project jointly sponsored by the Egyptian Ministry of Agriculture, University of California, and USAID. The views expressed are solely those of the authors. This article originally appeared as Giannini Foundation Paper 615.
1 Michael J. Piore, *Birds of Passage: Long-Distance Migrants in Industrial Societies* (New York: Cambridge University Press, 1978).

2 Charles P. Kindleberger, *Europe's Postwar Growth* (Cambridge, Mass.: Harvard University Press, 1967).
3 Jonathon Power, "Faulty Foundations for Europe's Growth." *New York Times* (5 February 1973).
4 W. R. Bohning, "International Migration in Western Europe: Reflections on the Past Five Years," *International Labour Review* 118, no. 4 (1979): 401–14.
5 J. S. Birks and C. A. Sinclair, *International Migration and Development in the Arab Region* (Geneva: International Labour Organisation, 1980).
6 Marshall Hodgson, *The Venture of Islam* (Chicago: University of Chicago Press, 1974), vol. 2, p. 117.

PART 4

Introduction
Geography and society

DOREEN MASSEY

The fact that 'Geography matters' has been an underlying argument of the whole of this book, but in each section we have treated it in a different way. In the first section we looked at the social significance of conceptualization and at its relationship to developments both within and between societies. In the second section we looked at the significance of 'geography' in the constitution and operation of a number of very different social processes. Our argument there, and throughout, has been not only that the geography of society is socially constructed and that, to understand it, that fact must be recognized, but also that social processes and phenomena are constituted geographically. The corollary, therefore, is that to understand them account must be taken of their geography. In that second section, we considered this proposition in relation to the organization of the city, the constitution and reproduction of cultural forms, and the operation of international law. In the third section we turned our attention to the central question of the construction of 'place' and of geographical variation within the wider system, whether that be international or national.

In this final section we tackle the question at the broadest level of all, the level which allows us to pull together all our arguments: why does geography, in the sense in which we have defined it, matter to the development of society as a whole? Let us take an example which builds upon the conclusions of the last section. Having constructed our syntheses, having recognized that the international world, or the individual nation state, is a set of spatially organized interdependencies, a mosaic of unique places – what does it matter? It matters because internal variation, and the mechanisms which that sets in train, can alter the development of the whole. The geography of a society is a fundamental component of how that society will reproduce itself, develop, and be changed.

For example, since the 1960s in the United Kingdom a whole range of well-recognized 'national' changes have been taking place. The British economy, and British manufacturing in particular, has been in a state of more or less unrelieved crisis. The occupational structure of employment has changed dramatically, and so has the social structure. Skilled manual jobs have declined in number, white-collar ones have increased, as has employment

161

in managerial and technical functions. The proportion of married women in the waged labour force has grown considerably (see Chapters 6 and 7). And with these changes in occupation have gone changes too in social structure; there has once again been talk of 'the end of the working class as we know it' (see Chapter 3). The level of employment in relation to the potentially economically active population has changed – unemployment, in other words, has increased beyond what was imaginable in the sixties. The industrial structure of the economy has shifted: both primary and manufacturing industries have declined in the proportion of the population they employ, while service industries have grown. And, of course, the governing politics have shifted too, from the technological-revolution-plus-reform of Harold Wilson to the right-wing radicalism of Margaret Thatcher.

Geographical structure and geographical reorganization have been central elements in all of this. For British industry, locational flexibility has been a way of lowering labour costs in the face of increasing international competition. Both internationally and nationally it has sought out reserves of cheaper and more vulnerable labour, in the Third World or in the less urban areas and the peripheral regions of Britain. That movement, together with the spatially uneven impact of the more general decline of manufacturing, has left behind it huge problems of dereliction and unemployment. The concentration of these in the inner cities and the responses which such a situation in part provoked have consistently been issues on the national political agenda. Whether it be urban riots, conflict between national and local government, or political clashes over what should be the form of metropolitan government, the big cities are a national political issue. The geographical unevenness of unemployment more generally has had other repercussions, too. The consistently high levels in the peripheral regions of declining heavy industry have claimed attention ever since the end of the long boom. The regional policy which on the surface was a response to geographical unevenness in fact played a complex role. Calming the anger of those in the regions on the one hand, it was also the lever used to enable further jobs to be lost as the nationalized industries were restructured (don't worry, if we close your mine/steelworks, we'll give you Special Development Area status and new jobs will arrive to take their place). But if regional policy was a response to the demands of some parts of labour, it was also, more widely, a means of undermining labour's strength. Wage bargaining had become led by the central regions and any spreading around of jobs (and also therefore of unemployment) also slackened that bargaining power. And the political function of regional policy was important in the sixties, for it was in the regions that was found the social base of the governing (Labour) party, and the demands of that highly geographically localized base had to be responded to. Over the two decades since then the geographical outlines of party-political bases have, if anything, become more entrenched. The national changes in occupational structure and social class have been taking place differentially between regions, and in certain ways have reinforced some of the old north–south divisions, if in a changed form. Indeed, the geography of social change may even have reinforced its social content. The 'new middle class' of the outer south-east may be clearer in its social distance from 'the working class' because of its very geographical separation from the heartlands of production. And that social divide in turn reinforces the political divide.

And so on; any thorough understanding of the enormous changes which have occurred in Britain over the last few decades must take account also of the geographical changes which have taken place. And this is not just because a knowledge of spatial change will add to the detail, put flesh on the bones. It is also because without the knowledge of the spatial change, some of the national processes will be very hard to understand. More generally, understanding society entails also understanding its geography. To do this it is necessary to pull together and build on the arguments in the central sections of the book. It is to this that this final section is addressed. And it is through two issues of central current significance that we have chosen to illustrate our thesis.

The increasing degree of international capitalist interdependence has been a theme throughout this Reader. In his chapter in this section, Martin Kolinsky looks at its impact upon one of the basic political entities in the geographical organization of the modern world – the nation-state. Robert Sack, in his contribution in the first section of the Reader, stressed the centrality of the nation-state, with its fundamentally spatial/territorial connotations, to the functioning of modern society. And yet, in some ways, both the centrality and the coherence of the nation-state might seem to be under threat – the centrality because of the proliferation of supranational organizations, and the coherence as a result of the growth of sub-national regionalisms (or nationalisms).

A whole range of causes (considerations of defence, of culture, of politics, of economics) lie behind these conflicting trends. Important among them, once again, is the fact of a shifting international division of labour. This changing geographical structure of the world economic order on the one hand tends to pull states together in an attempt to increase their leverage on wider forces (the relationship here to Picciotto's argument about law is clear), and on the other hand by reducing regional economies to incoherent branch-plant appendages of distant economic structures, tends to provoke regional nationalisms. The second direction of this tension, the tendency towards regional nationalism, recalls Clarke's arguments in an earlier section, that geographical cultural variations are by no means dead and that indeed the very disruption of accustomed patterns can itself lead to a nostalgia for cultural roots in place and region. What is at issue in this dual movement, above and below the nation-state, is precisely the tension between inter-dependence and uniqueness.

Within countries, Kolinsky's chapter brings out clearly the number of different ways in which internal geographical variation, in different forms, has been important in different countries. In Germany the fact that Franz Josef Strauss has such a strong regional base has had its effect on national politics. In the United Kingdom there has been both regional nationalism and the infinite (and changing) complexities of regionally based economic demands. In France there has been both nationalism and, very importantly, the decentralizing demands of regional economic planning. In Italy constitutional delays were caused by the problem (for the central government) of simultaneously responding to regional pressures and of keeping out of power a political party in opposition to the national ruling party. In Belgium a mixture of linguistic and cultural contrasts reinforced by highly unequal economic fortunes, and exacerbated by recent internal geographical movements, has at times severely disrupted national politics.

These issues are immediate and current, and they change fast. The economic strength of the regions of Belgium has reversed in the last decades. In the UK at the time of writing the West Midlands looks like getting some regional aid – a product not just of uneven economic decline but also, probably, of the changing political map. At a supranational level the need to coordinate is still made problematical by the internal diversity of interests and by national political assertion. The interplay between interdependence and uniqueness continues.

Kolinsky's article concerns the impact back on society of its spatial organization. But arguments apply at the same level to questions of the relationship between the social and the natural – the presently dominant form of that relationship currently threatens serious repercussions for society. It is this question which the article by Francis Sandbach addresses, in the final contribution to this collection.

Sandbach rejects the simplistic polarization of views which has sometimes characterized arguments on the issue of ecodoom: on the one hand the prophets of inevitable catastrophe, on the other those who pin their faith in market mechanisms and new technology as averters of disaster. What Sandbach explores is the far deeper argument of the relation between economic and social organization and resource–scarcity and pollution. If nature is hitting back in particular ways, he argues, it is because of the dominance of certain underlying principles of economic and social organization. And if we are to respond it is those principles of organization which must be addressed.

Some of the examples which Sandbach gives of alternatives to the presently dominant way recall the discussion in Chapter 1, by Gold, and insist again on the variety of ways in which it is possible to organize the relation between the social and the natural. A subsidiary aspect which emerges in this discussion (for instance in the example of China) is the importance of spatial structure itself in the relation between the social and the natural. If, in other words, it is not just development, but the *form* of development which is important in structuring the social/natural relation, then one aspect of the form of development is precisely its spatial form. Two of our themes weave together here.

This position, by stressing that many of our environmental problems are socially produced, does not thereby nullify the existence or impact of 'the Natural'. On the contrary, the problems are real and what they demand (the real substance of their impact) will – hopefully – not be ecodoom, but a change in the way we organize society.

9

The nation-state in western Europe: erosion from 'above' and 'below'?

MARTIN KOLINKSY*

The centralized nation-state was the outcome of political evolution. Whatever the merits of the nation-state, it became apparent in the twentieth century that it did not prevent the mounting horrors of war, and that alone it could not ensure economic prosperity. As a result, efforts were made to develop new political devices. The preferred solution among those of liberal outlook was the 'international institution'. The aftermath of the First World War saw an attempt at institutional order through the League of Nations; the close of the Second World War brought the United Nations and a range of other more specialized bodies, some almost worldwide, others of lesser scope.

In Europe, the cradle of the nation-state, the most advanced attempts to develop a new institutional order are to be found. The idea of the nation-state as an ultimate, compelling reality was brought into question in Western Europe by the Second World War more widely and profoundly than had been the case after the First World War. The governments that had fused extreme nationalism and dictatorship, Nazi Germany and fascist Italy, were buried in the war of aggressive brutality they had unleashed. International relations were restructured by alignments of states dominated by the new military superpowers. The battered nations of Europe were corralled into one or other of the two great blocs as world politics became dominated by the Cold War. In the West, the nation-states retained their sovereignty (or gradually attained it in the case of the Federal Republic of Germany), but had to recognize the limitations on its exercise given their military, economic and financial dependence on the United States of America. Moreover, the perception of the threat of totalitarian communism, soon after the harrowing struggle against Nazism, made closer coordination of policies desirable and necessary to protect the shared values of democracy and liberty. Although the notion of a politically united Europe proved to be a fragile flower, important international structures were established such as the Organisation for

* Source: L. Tivey (ed.), *The Nation-State* (Martin Robertson, Oxford, 1981), chapter 4, pp. 82–103.

Editors' note: Most of the references, the footnotes, and small parts of the original have been removed in this version of the chapter.

European Economic Cooperation (OEEC) in 1948, which was transformed in 1960 into the Organisation for Economic Co-operation and Development (OECD), the Western European Union (WEU) in 1954, the North Atlantic Treaty Organisation (NATO) in 1949, the European Coal and Steel Community (ECSC) in 1952, Euratom in 1958, and the European Economic Community (EEC) in 1958.

The various organizations differed in type, but represented permanent alliances for specific purposes of policy alignment and coordination. The real issue was not surrendering political sovereignty but ascertaining common goals and finding methods of working together. The outcome was that the operation of government in Western Europe was not merely determined negatively by dependence on the United States. It was also influenced in a positive manner by participation in the increasingly dense network of international organization. For the newly created state of Western Germany, participation in European integration was a means of acquiring respectability and legitimacy, both requisites for the removal of the Occupation Statute and for winning independent, sovereign status. For Italy and the smaller countries, participation in the European organizations also represented a gain in status and a strengthening of their political integrity. Participation meant a defined place in the international order, formal recognition of its voice and interests, security and material benefits and, most important of all, a basis for positive action and influence. Dependence on a large neighbouring state, as was later to be repeated in the case of Ireland vis-à-vis Great Britain, was reduced and replaced by an acknowledged, formal (proportionate) equality.

However, the situation appeared to be more complex to British and French governments. Under both Labour and Conservative administration, Britain was not interested in anything beyond multi-lateral, loosely structured cooperation and consequently remained aloof from the initial phases of European integration. However, the economic success of these institutions, and the search for a new political role in Europe, led to a reversal of policy by the leaders of both major parties in the 1960s. The situation in France was almost the opposite, with governments seeking military independence while accepting some degree of economic integration with neighbouring countries. Nevertheless in both cases the network of international organization created a new environment for governments and significantly modified the traditional notion of national sovereignty.

The changing notion of sovereignty

Sovereignty means that the state is the ultimate source of authority, law and legitimate force within its boundaries. However the boundaries are not impermeable, and no state can be taken in isolation: its industry, energy, raw materials, finance and trade are dependent on world circumstances; its society is penetrated by international trends and influences. The exercise of sovereignty is further conditioned and modified by membership of international organizations, the obligations incurred there and the concern for maintaining positive relations with a great variety of other states. In reality, therefore, sovereignty is a more diffuse and indeterminate notion than was understood when the British Empire flourished and European states could act as world powers. There are some who are concerned that the development of supranational integration in the European Community (EC) will lead to a complete transfer of powers, so that countries such as Britain will become reduced to the status of provinces. Such a surrender seems most unlikely, as is discussed later, because of the reassertion of the authority of the nation-states within the Community. The somewhat paradoxical consequence of 'integration' in the EC is that national governments dominate the decision-making and policy-making processes. The Commission, representing the supranational element in the Community, is relatively weak compared with the representatives of national governments (the Council of Ministers) and of national civil services (Committee of Permanent Representatives, COREPER). Moreover, national parliaments have for the most part not established effective methods of controlling ministers in their dealings at Brussels, so that the authority of national governments is enhanced by the executive-to-executive relationships created in the Community. The diffusion of sovereignty is accompanied by a strengthening of central government powers.

The traditional meaning of sovereignty in unitary states has also come into question from 'below' by moves towards regionalism, autonomy and federalism in various countries. Although it did not arise from sub-state nationalism, the most significant change occurred in Germany with the establishment of the Federal Republic after the fusion of the western zones of occupation. West German federalism, which arose from special circumstances, is discussed in the next section. While such a solution is advocated by only minorities elsewhere, pressures for regional reforms have been widespread, even in the staunchest of unitary states. The situations in Britain, France, Italy, Belgium and Spain are reviewed briefly below.

The causes of sub-state nationalism and regionalism are various, and each case has its unique aspects. Suffice it to state here, without attempting to elucidate causal explanations, that the pressures for decentralization, and in some cases for autonomy, arose from the overwhelming concentration of powers in central government at a time when the growing interdependence of states seemed to bring the capabilities and functions of the traditional structure into question. It is somewhat paradoxical that while interdependence reinforces the concentration of powers (because the action is at a government-to-government level), it represents at the same time limitations on traditional capabilities (in economic, military and diplomatic spheres), which in turn reinforces the questioning of the validity of such concentration of powers. In Britain, for example, the sense of political and economic decline, while leading to membership of the EC, also contributed to the erosion of some of the links between Scotland and England. It was in this context of change that the rise of Scottish nationalism stimulated in the 1970s intense discussion of the political alternatives of devolution, independence, and even federalism.

The changes occurring 'above' the level of the state, are not without consequence for the patterns of authority and political integration within national structures. There is posed the question of the redistribution of functions for the sake of greater administrative efficiency and democracy (as in the *Royal Commission on the Constitution* – the Kilbrandon Report, 1973): How to relieve the congestion of affairs at the centre?

How to promote regional development and planning to reduce socio-economic imbalances?

How to encourage democratic participation at sub-state levels?

Although internal decentralization provides greater scope for cultural and regional diversity, recent experience in Britain and France suggests strongly that unitary state governments are determined to maintain full control over whatever measures of change are introduced. The political circumstances affecting the trends of change in various countries are examined in the next section, which is followed by further consideration of the problem of interdependence and the various national interpretations and reactions to it.

Pressures for decentralization: sub-state nationalism and regionalism

Most of the constitutions of the Western European countries provide for a concentration of powers in central government. In contrast to the unitary states, there are only three examples of federalism: Switzerland, Austria and West Germany. The Federal Republic of Germany, established in 1949, consists of eleven *Länder* (including West Berlin). Legislative powers were distributed between the federal and the *Länder* governments. The former has exclusive power in foreign affairs, citizenship, currency and communications, while sharing powers (concurrent legislation) with the *Länder* in less vital matters. The powers of the *Länder* are considerable, including control of education, cultural affairs (radio and television) and police. Moreover, the implementation of federal legislation is largely accomplished through the *Länder*. Hence the importance of the *Bundesrat*, the second parliamentary chamber in which each *Land* is represented according to its size. The *Bundesrat*, which has a suspensory veto on bills passed in the *Bundestag*, serves as a meeting-place for officials of the central government and of the *Länder*. With the exception of Bavaria, regionalist feeling is not pronounced, and there are constitutional provisions to even out regional discrepancies. In conformity with the aim of 'unity of living standards' in all *Länder*, the federal government redistributes certain taxes in favour of the poorer states. In Bavaria, with its special historical and political traditions, there is a strong sense of particularism. Although the Christian Social Union (CSU) is a permanent coalition partner of the Christian Democratic Party (CDU), it is by no means passively acquiescent, and there have been threats to form a separate national party of the ultra-conservative right. The political ethos of the CSU is notably on the right-wing of Christian democracy and its powerful leader Franz-Josef Strauss has led the opposition on such important issues as the *Ostpolitik*. Although his ambitions to become the Chancellor-candidate were long frustrated, he was sufficiently influential to undermine two CDU leaders (Barzel and Kohl) and in June 1979 finally reached the top position in the Christian Democratic coalition. It is his extremely powerful base in Bavaria, the second largest state of some eleven millions, that has enabled Strauss to stay for so long in the forefront of national politics.

Although the decentralization of power involves elected parliamentary forums (the *Landtage*), there has been a constant decline in their legislative power and a growth in the bureaucratic coordination of federal/*Länder* relations. Kurt Sontheimer has summarized the process in the following words:

The practice of co-ordination and co-operation in German federalism which is supported by innumerable treaties and administrative agreements between the Laender and between the Federation and the Laender is carried out mainly without the Laender parliaments. Federalism as it is practised is to a great extent a matter for bureaucrats, not for politicians, and it has withdrawn in part from parliamentary control. (Sontheimer, 1972, p. 154)

The tendency, therefore, is towards a certain degree of *recentralization* of power with a strong element of administrative devolution. The main actors on the *Land* level are the minister–presidents and officials of the *Länder* ministries. The tendency towards recentralization has been reinforced by several other trends. The problem of terrorism, for example, has necessitated much closer coordination among the *Länder* police authorities and federal security bodies. The issues of university and educational reforms have also prompted efforts at greater coordination, as have economic questions. In all these areas, the federal government has taken initiatives both for closer cooperation with the *Länder* authorities and to establish greater control for itself. Indeed, with the internationalization of European economies becoming ever more pronounced – within the European Community and in terms of dependence on world trade – the need for consistent *national* economic policy is keenly felt in Germany. Moreover the huge success of the German economy, and the financial transfers reducing regional discrepancies, have largely overshadowed such problems as the growing obsolescence of much industry in the Ruhr area. Such decline is offset by the extremely rapid development of *Länder* like Baden-Wurttemberg, which used to send its surplus labour across the border to work in Alsace, but by the mid-1960s reversed the direction of migration as its industries grew to contribute nearly one-sixth of the Federal Republic's gross national product.

The trends toward recentralization have weakened the parliamentary structures of the *Länder*, but have not undermined the federal system as such. The federal institutions (*Bundesrat*, *Länder* executives, Constitutional Court, and the constitution itself) remain viable, and political party organization is well adapted to the federal organization of the state. Unlike the situation of the Weimar Republic the legitimacy of the political system is not in question. Nevertheless it is clear that federalism as practised in West Germany is not static, and there is a movement towards a degree of recentralization. This contrasts with the situation in Canada, where there is a pronounced shift to the provinces (Quebec nationalism, Alberta economic strength). It contrasts also with the situations in which many European unitary states have found themselves, where the pressures for decentralization have been often strong enough to pose the question of altering the established constitutional order. These pressures are strongest when reinforced by nationalist claims (Scotland, Corsica, the Basque country, and in Belgium) though regional feeling based on aspirations for economic development are also prevalent in such diverse areas as the west of France, the south of Italy and the north of England.

The rise of nationalism in Scotland and Wales resulted first in the *Royal Commission on the Constitution 1969–73* (Kilbrandon Report) and then in legislative proposals for devolution. The intense and prolonged debates on these proposals were marked by backbench revolts that seriously embarrassed the Labour government and contributed to its final downfall in March 1979. Considerable government activity during the 1970s was focused on the implications of the Kilbrandon Report. In addition to working out the extremely complex devolution proposals, the cabinet and cabinet office committees were concerned with mounting regional economic pressures. Development agencies were established in Scotland, Wales and Northern Ireland, and regional policy generally became much more active. Grants for industrial location and other aids under the Wilson/Callaghan administration of 1974–9 rose to over £400 million a year.

Inevitably the effect of government attention to the Celtic peripheries, especially to Scotland, created concern in the less affluent English regions. It was most pronounced in the north of England, which felt that it was in as much need of assistance as Scotland but had less political influence on the government. Not surprisingly, then, Labour MPs from the North East joined the backbench rebels in voting against key aspects of the government's devolution programme, and lobbied for the establishment of a Northeast development agency. The repercussions were felt further afield. Although the West Midlands was not as directly worried by the devolution proposals, the sharp decline in the prosperity of the region led to growing resentment at not being included in the government's category of assisted areas. The Birmingham Chamber of Commerce, for example, has consistently argued that the city has lost its growth industries by years of government direction of investment away from the industrial heartland. Moreover, it cannot even tap the European Regional Development Fund, because Westminster did not include it among the assisted areas. The resentment is underlined by a feeling of political weakness: unlike traditionally poor areas the organization of a regional interest grouping through one of the major parties had not been necessary in the past. However, in its situation of declining prosperity the lack of a special channel to government bounty was keenly felt in the West Midlands. These were the reactions of people concerned with industrial and planning problems, but they were not more widely articulated into mass-based political demands. Since political expressions of regional identity did not emerge in other parts of England either, there were no parallels with the Scottish and Welsh nationalist movements. For the latter, devolution was seen only as a step along the way to independence, despite the fact that in the government's view devolution was set out as a means of responding to aspirations for democratic participation at an intermediate (regional) level of government. In fact this, rather than national independence, was consistently the most popular option in Scottish opinion polls. Despite the scrapping of the Labour devolution bill, the question of change in the political and administrative relationships of the constituent parts of the United Kingdom remains open. Whatever the future outcome, however, it seems unlikely that decentralization measures will reduce the essential controlling and mediating capacities of central government in Westminster. This expectation is supported by the French experience, where the 1972 regional reforms were by no means as far-reaching in their implications as the changes envisaged in the Scotland Bill.

Unlike Britain, French government concern was impelled less by nationalist movements than by a need for a new administrative framework for promoting regional economic development. The administrative reform was limited because the intention was not to replace the existing pattern of *départements*, established by Napoleon, but to bring them into association for planning purposes. The potential political significance of the regions was curtailed by rejecting direct elections. Instead the regional councils represent for the most part the established political interests of the parties and the local notables. Nevertheless the *départements* are too small on their own, given the urban–industrial development of France and the economic pressures exerted on it by Common Market competition. The relatively simple solution sought in the 1972 reforms – the coexistence of regions and *départements* – could lead to greater tensions if nationalist or regionalist feeling should develop more strongly, perhaps in response to prolonged recession, in sensitive areas such as Brittany, the midi, or Corsica. Even if that does not happen, it has become increasingly clear that strong-minded regional elites, such as those in the Pays de la Loire and in the Lyon area, want more scope for themselves and less constraint from Paris.

In Italy, the situation is dominated by the division between the industrial North and the underdeveloped South (the *Mezzogiorno*), a problem recognized in the Treaty of Rome, which established the European Economic Community, as calling for special measures of assistance. These have come from both the European Investment Bank and the European Regional Development Fund. Despite these actions, which have conjoined with three decades of Italian government development projects, the Mezzogiorno remains a special problem, whether measured in terms of its low contribution to gross national product, its poor living standards or its high rate of unemployment. The regional disparities in Italy have remained the most pronounced in the European Community, and continue to impede Italy's progress toward greater economic stability. As the Community has recognized from the beginning, it is not merely an Italian problem but is a serious threat to the aim of creating integration, which requires greater similarity in economic performance, standards of living and social opportunities.

Constitutionally Italy is a unitary state with regional administrations. As a reaction against fascism, the 1947 constitution provided for regions as an integral part of the organization of the state. But the establishment of the regions was held back until 1970 because of the fear on the part of Christian Democratic governments of Communist domination in the central regions of Emilia, Tuscany and Umbria. However, the shift to centre–left coalitions in the 1960s finally resulted in the introduction of the regions because of the insistence of the Socialists when invited to join the governing coalition. There are twenty regions, each with a council elected by proportional representation as in national elections. The council elects its executive and presiding officer. The legislative powers include agriculture, health services, planning and cultural affairs – a range considerably less than those of the German *Länder*, or even what was contemplated in the Scotland Bill, though Italian regions do have limited powers for raising revenue. Five of the regions, including the islands, are defined as special and have somewhat wider legislative powers. An area of particular concern for all regions is industrial policy, which is determined by central government, although the regions are responsible for small business. The division is a cause of friction because the regions argue that it is impossible to establish development plans without being able to influence industrial policy. Nevertheless, dynamic regions such as Piedmont have approved regional plans.

The transfer of powers to the regional administrations has proved to be a slow and uneven process. As may be expected, those institutions, such as the employment office, that are important sources of political patronage have proved much harder to decentralize than less vital administrations. Another problem is financial. Regional budgets are mainly financed from a common fund (i.e. nationally raised revenue) distributed by central government. A frequently voiced complaint is that Rome controls the purse-strings too tightly. But there is some evidence that certain regions, particularly in the Mezzogiorno, are seriously underspending their allocations, either because of inefficiency or because of lack of programmes. However, these problems have to be seen in perspective: the slow pace of devolving powers has meant that operative regional government is a very recent phenomenon in Italy. It is far too early to attempt to judge its effectiveness in the urgent tasks of development, planning and administrative reform. It is not made easier by the frequency and apparent intractability of national government crises.

Belgium is another example of a unitary state with pronounced regional differences. The most prominent aspect is a linguistic/cultural conflict between Flanders and French-speaking Wallonia, which has persisted over decades, and which has more recently developed further into a dispute about the status of Brussels. Over the past twenty years, with the decline of traditional steel and mining industries, economic dominance has shifted from Wallonia to the north, which has prospered on American and German investment (the port of Antwerp in particular; Flanders generally because of its less strike-prone and less organized labour force). Whereas the Walloons have reacted by demanding priority action to revitalize their obsolescent industries, the Dutch-speaking Flemings who form a majority of the population (5.3 out of 8.8 million) resent the cultural and political domination of the French-speaking elite and are determined to preserve their newly acquired economic advantages. The long simmering conflict increasingly focused on the question of Brussels, a predominantly French-speaking city in the southern part of Flanders. The steady shift of population from the city centre to the suburbs has created politically significant pockets of French speakers. The reaction of the Flemish communes has been strong. The government of Leo Tindemans (1977–8) attempted to resolve the tensions with a programme of regional reform that proposed the replacement of the nine provincial administrations by the three regions of Flanders, Wallonia and Brussels, with a number of central government powers to be transferred to elected regional and sub-regional councils. Although it had been agreed that Brussels' boundaries would remain within the limits of its existing nineteen communes, the prolonged negotiations over the regionalization legislation were seriously interrupted in 1978 by objections from the Flemish wing of Tindemans' own party (the Social Christians, CVP). Tindemans resigned, but the subsequent elections proved inconclusive, and the stalemate over constitutional reforms continued to the end of the decade.

The irony of the situation is that until his resignation Tindemans was seen in the larger EC context as one of the chief movers of European integration. His sober and cautious report to the European Council, 'European Union' (January 1976), was widely discussed and, though criticized in detail, regarded as heralding a more realistic and useful approach to the problem of European unity. A more fundamental irony is that Brussels, the seat of the Community, has become the centre of conflict over national unity.

The enlargement of the Community to include Spain is likely further to intensify regional problems. First, there is the pressure of Spanish agriculture on the poorer farming regions of France and Italy, in mutual competition over Mediterranean products of cheap wines, and fruit and vegetables. This created strong feeling in southern France and Italy once the prospect of Spain's entry became a reality. Secondly, the poor regions of Spain, such as Andalucia and Extramadura, like the Mezzogiorno, greatly intensify regional disparities within the Community and will strain the limited resources of the Regional and Social funds. Thirdly, within Spain itself, the long-standing violence of the Basque situation threatens national unity and the fragile democratic order. Basque and Catalan autonomy were ruthlessly suppressed during the Franco dictatorship, but in the new constitution of 1978 the right to autonomy of the 'regions and nationalities' of Spain is recognized and guaranteed (article 2). However, somewhat as in Italy, fulfilment has taken longer than expected, and the question of defining the extent of devolution is subject to important political reservations.

Central government control

The trends in the various countries should be seen in the fluid international context. The growth of the world economy, the energy crisis, and the claims of the Third World during the 1970s – notably in the United Nations Committee on Trade and Development (UNCTAD), General Agreement on Tariffs and Trade (GATT) and Lomé Convention negotiations – have contributed towards what has been described as 'the new nationalism' among Western countries. The effect of the 'new nationalism' is to enhance the importance of central government control, which, as is discussed below, is being reinforced by the process of European integration. However, integration carries with it a certain counter-pressure in regions on the periphery or in economic decline to acquire more say in the management of their problems. This may be fuelled by invidious comparisons: the feeling in North East England of coming off second-best to Scotland in benefits and influence; the resentment in the West Midlands at being excluded from regional aids; the feeling in the West of France that not enough is being done to speed decentralization of Paris industries and administrations; the sense of neglect in Alsace, which tends to measure itself against the standards achieved across the border in Baden-Wuerttemberg and Switzerland. It is true that these responses can be managed by central government through generous policies of regional aid and through political payoffs where appropriate channels have been established. But there remains the question of democratic participation in decisions affecting the regions. For example, the seventy-one Scottish MPs in London are not well-placed to exercise control over the Scottish civil service in St Andrews House, Edinburgh. Nor has regional reform in France represented much improvement because the councils are not directly elected, and the government has strongly resisted pressures to make them more viable political institutions. However, as the case of West Germany illustrates, elected assemblies in themselves are not the full answer to the problem of democratic participation. The decline in the legislative powers of the *Landtage* and the tendency to bureaucratize regional affairs indicate the general difficulty that parliaments have become relatively marginal to decision-making processes in many spheres. The multiplication of regional assemblies in itself will not reduce the possibility that such institutions may be emptied of much of their content and lose their potential influence. The problem is also posed decisively at the level of the European Parliament, as will be considered later.

Pressures for decentralization affect the exercise of state power in that political stability requires attention to the demands of sensitive regions. But the central administration alone has the capacity and information to manage the competition for limited resources and to exercise overall responsibility. Therefore, although its authority may be questioned internally, and its means in the world reduced, the political grip of the unitary state has not faltered, nor is it likely to do so.

International organization

The nation-states of Western Europe participate in numerous international organizations, but the European Community is the only structure that embodies the principle of supranational integration. The interpretation of the supranational aim varies from federalism – that is, the creation of a United States of Europe on the model of the USA, with transfer of sovereignty in the fullest sense to a European government – to a permanent association of independent states, retaining their capacity to decide their domestic and foreign policies for themselves, but seeking common ground with their partners by means of consultation and coordination. Whereas very few partisans of federalism remain after the hard realities of de Gaulle's nationalism in the 1960s and the energy crises of the 1970s, the EC has survived as an association of nation-states. Its importance in world trade and economic affairs is not in doubt; it has elaborated a unique legal framework through its treaties and the Court of Justice; and it has strengthened the potential of the European Parliament as a parliament-in-making through the introduction of direct elections. While these supranational elements serve to buttress the Community framework, the limitations have to be stressed as well. The European Parliament is a consultative body and its restricted powers cannot be extended without the agreement of all the governments and parliaments throughout the Community. The Court of Justice has only theoretical superiority over national institutions and does not have means of enforcing its decisions. Of greater importance is the fact that the Commission, which embodies the supranational animus of the Community, lost much of its momentum after its collision with de Gaulle in the mid-1960s. The influence of the national governments, which is exercised through the powerful Council of Ministers and the Committee of Permanent Representatives, has been enhanced by regular summit meetings of prime ministers and presidents. At the same time, the power of central governments has increased domestically because the intergovernmental bargaining, and the legislation arising from it, is very difficult for parliaments to monitor and control.

Despite these limitations, which arise from a tangle of cross-purposes and competitive bargaining about the terms of cooperation, the Community has proved its capacity to survive the shattering of illusions about the progress and prospects of supranational integration. What has endured is the underlying interdependence of the member states and the importance of their multilateral contacts, even in a less than ideal organization. [...] A major theme used to justify the Community is that it is a safeguard of democracy against the threat of a recurrence of totalitarianism (in its most doggerel form it portrays the Second World War as the 'European civil war'!).

The truth, of course, is that the EC by itself has nowhere near the cohesion to affect the basic political structures and trends within the member states. [...] Nevertheless, the Community represents an important aspiration that has imparted a wider sense of identity and belonging, especially to the post-war generations and it is a source of pressure on applicant countries for upholding democratic norms. The further enlargement of the Community to include Greece, Spain and Portugal is a significant test. With luck, the Community may accomplish what the military alliance was unable to do, namely to underpin and strengthen the new, fragile democratic structures of those countries by facilitating their economic development.

The existence of the Community, with its processes of political and economic concentration, contributes in some degree to the cohesion of the wider NATO alliance of fifteen countries. But it is indirect, a spin-off from the multiplicity of contact, rather than an indication of the complementarity of the two organizations. NATO, which is an intergovernmental structure, represents a pooling of common defence and strategic interests under American leadership. In this perspective the EC bloc dissolves into an amorphous assembly of states grouped around a superpower. Even France, which has withdrawn militarily, has consistently recognized its ultimate reliance on American nuclear protection by remaining within the NATO political alliance. [...]

National interpretations of interdependence

It is evident that notwithstanding the viability of the military alliance, the political community underlying it has not progressed beyond embryonic stages and remains in a state of flux. Economic discussions in various international forums and political consultation in the EEC (a grouping narrower than NATO) partially supplements it, but not sufficiently to create an integral system of military – economic – political – diplomatic coordination. The instability or low-level crisis arises from the unbalanced nature of the Atlantic alliance, as well as from the varying bilateral and multilateral interests of the European partner states. The ambiguity was well characterized by Hanrieder (1979) as 'tendencies towards divergence and tendencies toward integration'. Ambiguity is also pronounced within the grouping of EC states that are most similar in size. After two decades of cooperation, the degree of supranational integration in the EC remains minimal, though consultations have muffled all-out divergencies in some economic fields. Most of the governments recognize that interdependence compels cooperative relations; but these are much easier to achieve in periods of economic growth than in situations of instability and insecurity. While economic vulnerability empha- sizes the underlying interdependence, it raises postures of nationalistic self-defence because national autonomy is, [according to Morse (1979, p. 66)], 'the most secure framework for control'. No state gives all: its general willingness to coordinate policies is tempered by the reserve of seeking competitive advantage where possible and by its anxiety to minimize its vulnerability to unpredictable changes in the external world. A finger is given, the hand held back.

When some grand vision emerges from an American or a French president, the kind of political consultation it inspires among the partner governments is, not surprisingly, accompanied by scepticism and enquiry into motives.

From the point of view of the individual state, membership in the various organizations is valued for the dual purposes already considered, that is, coordination of policies and protection of one's own interests. The option of falling back on national autonomy (at least in certain spheres) is always possible because an important limiting characteristic of international organ- izations is their lack of sanctions over member states. Thus France withdrew its territory and military forces from NATO, but chose to remain a member of the political alliance. There were no sanctions that could be applied to modify France's behaviour. Sanctions may be applied against a smaller state, but often with uncertain results. Turkey, for example, was not deterred by arms sanctions after its invasion and occupation of Cyprus, and the American ban was eventually lifted despite Greece's strong protests. Sanctions are usually limited in value because international organizations are associations of common interests, and it is difficult to apply sanctions without inflicting deep wounds on the organization itself. The experience is not confined to NATO; the agreed rules of the International Monetary Fund (IMF), General Agreement on Tariffs and Trade (GATT), International Energy Agency (IEA), European Community (EC), as well as of a host of lesser organizations, have been breached on many occasions without serious consequences for the undisciplined member states.

It is advantageous for the individual states to belong to several organizations, each specializing in a policy field, rather than to belong to one large political community that would permanently curtail the autonomy of the member states and perhaps reduce them to the level of provinces. In fear of such restraints Britain chose to stand clear of the early phases of European integration in the 1950s, while participating fully in the western military alliance. France reversed that choice in the 1960s and 1970s. Hence the idea of a Western European political bloc gradually achieving equality with the United States was severely unrealistic. It was not possible either to consider incorporating defence concerns into the EC or to achieve common policies in sensitive matters such as energy and foreign affairs. In any case, most of the issues stretch across the Atlantic, and usually across the Pacific too. While it is all too obvious that defence is inconceivable without the USA, it is no less true that the domestic and foreign economic policies of America are vital determinants of European financial, economic and energy problems too. In these spheres, bilateral interest relations (e.g. USA–Germany, USA–Britain) have not in the slightest diminished in importance. These are among the many reasons why the EC is so very far removed from becoming a European government, which implies *one* executive/legislative authority, *one* currency, *one* army and, above all, *one* sense of political community that surpasses the established orders of Britain, France, Germany, Italy, and the rest.

In fact, the diversity of international organizations corresponds with the *diversity of interests* of the nation-states and with the *scope of those interests*, ranging far beyond Western Europe to the other continents. The game has changed since the Second World War in that the conflict of national interests, backed by armed forces, no longer rules. The game has become that of seeking competitive advantage within coordinating organizations. Instead of old-fashioned trench warfare and civilian destruction, the present order is typified by bloody noses on the faces of apparently bloodless civil servants on away days in Brussels, and by ministerial managed-smiles after all-night conferences. Boring, but a decided improvement.

The game is not without its paradoxes and dangers. The international networks (including those with supranational elements) represent little more than extensions of the existing states. Policy formulation, as well as imple-mentation, still derives in the main from the central governments. The discussions are on a government-to-government level for the purpose of seeking coordination rather than to attempt common policies. Therefore, in its role as coordinator and mediator of policies at both national and transnational levels, central government remains in a key strategic position. The diffusion of its sovereignty rather paradoxically serves to strengthen its power of decision. Similarly at the sub-state level, where regional units exist in non-federal systems, the same tendency prevails. The central administration alone possesses the capacity and information necessary to oversee the entire scope of the governmental process. A further paradox in the game is that the introduction of additional democratic institutions (regional assemblies, the directly elected European Parliament) does not necessarily lead to more democracy. Since the process of decision-making has become so firmly on an executive-to-executive basis, involving primarily ministers, higher civil servants and representatives of organized interest groups, parliamentary bodies have suffered a long-term decline in their relevance. Parliaments may not be marginal institutions, but they are at a remove from the centres of policy formulation and implementation. What occurs in the national context is all the more likely to be repeated in newly emergent contexts. Where regional assemblies exist, their powers are narrowly circumscribed by central government, which moreover jealously protects its traditional prerogative of exclusive right of representation abroad (i.e. prohibits formal relations between a region and the Commission). The directly elected European Parliament is less easily manageable, but is handicapped by restricted powers and has to operate in an ill-defined situation of an embryonic political community. These paradoxes give rise to the danger that national parliaments, already weakened by a reduced capacity to control and scrutinize legislation and budgets, are forced to cope with complex systems of decision-making in which responsibility is much more elusive than the traditional doctrine of ministerial responsibility would suggest. The loss of capacity is not compensated by transfer of powers to the European Parliament or to regional assemblies. The emergent situation of new circuits of consultation and decision-making increases parliamentary weakness because power lost on the national level does not accrue elsewhere. It simply disappears, leaving central government with its bureaucracy freer of constraint. The danger is by no means that of impending dictatorship, but the ease with which responsibilities may be blurred. Public uncertainty as to where both power and responsibility lie can diminish the vitality of democracy if more effective controls over central government activity, appropriate to the changing circumstances, are not found.

References

Wolfram F. Hanrieder, 1979. 'Co-ordinating foreign policies' in Werner Link and Werner J. Feld (eds.) *The New Nationalism* (New York/Oxford: Pergamon).

Edward L. Morse, 1979. 'The new economic nationalism and the coordination of economic policies' in Link and Feld, *op. cit.*

Royal Commission on the Constitution 1969–1973 (Kilbrandon Report) 2 vols, Cmnd 5460 (October 1973).

Kurt Sontheimer, 1972. *The Government and Politics of West Germany* (London: Hutchinson).

10

Environmental futures

FRANCIS SANDBACH*

The question of whether or not there will be substantial physical and social limits to economic growth cannot be answered with any degree of certainty. While one can be certain that spring will follow winter, and autumn will follow summer, the same kind of certainty does not exist when forecasting the state of the environment in the future. The type of demands upon the environment will depend as today upon the form and extent of social activity. It is necessary to state the obvious, namely that different societies both today and in the past have made different demands and impacts upon the environment. If, for example, present-day energy consumption in different countries is compared, one tends to find that high levels of consumption occur in countries with high levels of economic activity. Nonetheless, there is a good deal of variation between countries with similar levels of economic activity. G. Foley comments:

Although Swedes and Canadians have roughly the same *per capita* GDP, Canadians consume on average twice as much energy. West Germany and the UK, on the other hand, have almost identical average energy consumption but the *per capita* GDP in West Germany is over 70 per cent higher than in the UK. (1976, p. 89)

Energy demands in the future will depend to a considerable extent upon the form of economic development. Using energy-accounting techniques, P. Chapman, G. Leach and others have demonstrated vast differences in energy consumption for different ways of producing similar objectives in transport, agriculture, heating houses, packaging of goods, etc. (see Chapman, 1975; Leach, 1976). For example, changes in the transport policy could have significant effects on energy demand.[...] However, environmental effects depend not only upon the choice of technology, but upon how it is organized. Hence, the policy of decentralization and local self-sufficiency in China has reduced the need for freight transport in general.

Changes in technology policy and forms of social organization could have

* Source: *Environment Ideology and Policy*, Basil Blackwell, 1980, pp. 200–23.

Editors' note: In reducing the length of the original, we have deleted many of Sandbach's sources as well as some data and some elaborations of his arguments. Readers interested in these should consult the original book.

marked influences on future demands upon the environment without necessarily affecting the level of economic activity, as conventionally measured in terms of GDP or GNP. Although it is stating the obvious to say that the future predicament of society is dependent upon the type of economic activity and planning that arises, it is nonetheless important to do so. There has been a stubborn belief that the future can be predicted by projecting, with little modification, trends from the past.[...]The question of whether such projections along the same path are desirable is seldom asked.

Those concerned with promoting alternative technology have, with some success, despite the limitations of political strategy, emphasized the possibility of alternative futures. There need not, for example, be an energy gap between demand for and supply of energy in the year 2000 if no nuclear power development takes place. Development of alternative technologies and conservation of energy could enable a better future to exist. It makes sense to consider various possible courses of action. These courses of action must, however, be grounded in political reality rather than Utopian blueprints. Alternative policies cannot be divorced from the social commitment necessary for their fruition. In the second part of this chapter, various policies will be considered in terms of their plausibility.

An assessment of future possibilities clearly depends upon an understanding of the real and imagined physical and social constraints upon economic development. So before discussing the merits of different policies, it is necessary to establish the scientific and ideological aspects of the varying opinions on the physical and social limits to growth. The 'limits to growth' debate is not of recent origin; many of its intellectual roots can be traced back to some of the principal economists and writers of the eighteenth and nineteenth centuries. The views of Thomas Malthus, David Ricardo, John Stuart Mill, W. Stanley Jevons, Karl Marx and Friedrich Engels were all concerned at one time or another with essentially the same issues. The ideas are not new, nor indeed are many of their shortcomings.

The limits to growth debate

Scientific and ideological views on the limits to growth debate can be broadly divided into three categories. In political terms they correspond to conservative, liberal and radical views, but they are couched in theoretical language. The first standpoint holds that there are physical and social limits of immediate concern. Physical limits are supported by the neo-Malthusian argument about exponential growth in a finite world, and the 'diminishing returns' hypothesis of Ricardo and Jevons. Social limits are supported by arguments concerning the social strain caused by growth and the depletion of positional goods, a view which can also be traced back to the writings of Ricardo, but which have been popularized recently in the work of F. Hirsch.

The second and more optimistic outlook is the economic/technological fix position of liberal economists. This stresses the mechanistic and economic responses to resource scarcity and pollution in a market economy. The third view is the Marxist political economy position. This stresses the interdependence of, on the one hand, the social organization of production and consumption and, on the other, the institutional superstructure that develops to control resource flow through the economy and to bring instruments of pollution control into action. According to this view, resource scarcity and pollution depend upon underlying principles of economic and social organization.

The neo-Malthusian position

In *An Essay on the Principle of Population*, Malthus argued for the existence of a universal law governing the relations between population and resource scarcity. Malthus argued that unless there are checks on population growth, the problems of subsistence itself will constrain population – the reason being that "population, when unchecked, increases in a geometrical ratio. Subsistence increases only in an arithmetical ratio" (1970; p. 71).

Malthus used historical data of population growth in the United States and Europe to give empirical justification to the geometrical increase in population. Without checks from lack of food or "peculiar causes of premature mortality", Malthus argues that the natural rate of population increases involved a doubling of the population every twenty-five years. Malthus might possibly be excused for assuming a cast-iron law of exponential growth, as demographic sources of data have been notoriously weak until much more recent times. Today, of course, it is known that improved standards of living and health, resulting in the main from economic growth, have been responsible for a decline in population growth rates in the world's developed nations. However, Malthus gave much less satisfactory evidence for the mere arithmetic growth in subsistence. It was this part of his theory that was attacked most strongly by nineteenth-century critics. Owenites refuted the assumption on the basis that if the soil was properly managed, vast populations could be supported.

Engels (1844) also took issue with Malthus, claiming that he had offered no proof that the productivity of the land could only increase arithmetically. Engels argued that, on the contrary, if population grew exponentially so too would the labour power employed to produce food. Furthermore, he went on:

If we assume that the increase of output associated with this increase of labour is not always proportionate to the latter, there still remains a third element – which the economists, however, never consider as important – namely science, the progress of which is just as limitless and at least as rapid as that of population (quoted from Meek, 1971; p. 63)

In successive editions of his *Essay*, Malthus came to rely increasingly upon the law of diminishing returns more usually associated with Ricardo. Ricardo's argument with respect to agriculture was that, as population grew, inferior land would have to be used and hence a lowering of productivity would follow. Jevons, in *The Coal Question* (first published 1865), extended the argument to the debate on coal resources. His Malthusian contemporaries argued that with the rate of coal consumption doubling every twenty years, coal reserves would be depleted by the year 2034. Jevons, following the diminishing returns argument, claimed that long before this the cost of fuel would rise as it became harder to mine and as demand outstripped supply. As Jevons thought that replacement by wind, geothermal or oil power was totally improbable, it was clear to him that progress was unlikely to be sustained.

This refined pessimistic view deserves close attention, especially with respect to physical resources, for many experts now appear to agree that there are not likely to be any constraints of an absolute type for many years to come. The amount of most resources in the first mile of the earth's crust probably exceeds presently known reserves by multiples ranging from thousands to millions (see Connelly and Perlman, 1975). And, contrary to the view that resources would become harder to extract, there is every indication that the opposite has occurred. Systematic studies by H. J. Barnett and C. Morse (1965) of the trends of extraction costs over the period 1870–1957 indicated without exception (apart from forestry and possibly copper), that the costs of exploitation had fallen significantly. Engels had been correct to assume the importance of science, for these changes in exploitation costs can be accounted for by technical change and substitution. To give a specific example: in the United States, the amount of energy needed to generate a kilowatt hour of electricity fell by just over 35 per cent between 1948 and 1968; and in Great Britain, the energy demand to produce a ton of steel decreased by 74 per cent between 1962 and 1972, chiefly as a consequence of the introduction of the basic oxygen process. [...]

In the 1960s, the Malthusian view of population pressure on limited resources regained its popularity. According to P. R. and A. H. Ehrlich (1970), if the population were to remain at 3.3 billion, then at current levels of demand lead would run out in 1983, platinum in 1984, uranium in 1990, oil in 2000, iron in 2375, coal in 2800, and so on. In *The Limits to Growth* study by D. H. Meadows *et al.* (1972), Malthusian assumptions lie behind the results of computer predictions. Data from 1900 to 1950 on the exponential growth of materials use, population, pollution, and the like are projected into a future which can only provide at best arithmetical growth in solving the problems of physical constraints. It needed no computer to follow the implications of the assumptions. As with earlier Malthusian predictions, the argument breaks down on empirical grounds because of the failure to allow for the expansion of resources to meet demand. The postulate concerning the fixity of exploitable resources has been proved wrong by scientific and technological developments as well as by increased exploration of surface minerals using currently known techniques. Moreover, as H. Kahn, W. Brown and L. Martel (1976) argue, there is little incentive to search for more reserves than would meet a few decades of demand. There would be little return on such investment and it might even be counter-productive as more known reserves could put pressure on current prices.

A few examples will help to illustrate how known exploitable reserves have increased in pace with industrial demand. For instance, in 1944 the United States prepared a study of its own known reserves of forty-one commodities. Had the predicted reserves remained static, then twenty-one of these commodities would now be exhausted (see Pehrson, 1945; and Page, 1973). Or take the example of aluminium: between 1941 and 1953 the known world bauxite reserves increased by an average of 50 million tons a year, and between 1950 and 1958 the average annual increase was about 250 million tons (Page, 1973). Or if one were to take the current concern for oil reserves: in 1938 the known reserves were sufficient for fifteen years' use at contemporary rates of consumption; in the early 1950s, after a doubling of consumption rates, the known reserves were sufficient for twenty-five years; in 1972, after a further trebling of consumption, the known reserves were sufficient for thirty-five years.

A useful distinction may be made between absolute resources in the earth's crust, exploitable resources, and currently known commercial reserves. According to this distinction, Zambia is rich in copper ores, but these are uneconomical to work under present conditions. Hence, Zambia has no copper reserves but plenty of copper resources (see Roberts, 1978). The weakness of neo-Malthusian arguments, as in *The Limits to Growth*, is that they often talk about reserves as if they were total resources.

A. Shenfield, former economic director of the Confederation of British Industry, dismisses the energy doom-mongers on three grounds. First, resources are assumed to be fixed; secondly, technological innovation is disregarded; and finally, predictions of doom in this field have so far proved to be false:

Thus in 1866 the United States Revenue Commission urged the development of synthetic fuels against the day in the 1890s when petroleum would be played out. In 1891 the United States Geological Survey declared that there was little or no oil in Texas. In 1914 the United States Bureau of Mines estimated that output would be six billion barrels in the whole remaining history of the country. This is now produced about every eighteen months. (1977, p. 16)

Neo-Malthusian arguments also pervaded the debate over pollution control. *The Limits to Growth* claimed that 'virtually every pollutant that has been measured as a function of time appears to be increasing exponentially' (Meadows *et al.*, 1972, p. 135). However, the dangers of projecting present growth rates exponentially into the future are well known. For example, it has been suggested that the trends of the 1880s might have shown cities of the 1970s buried under horse manure (Du Boff, 1974).

In *The Limits to Growth*, empirical justification for exponential growth in pollution is suggested from studies indicating rising levels of carbon dioxide, waste heat generation in the Los Angeles basin, nuclear wastes, oxygen content of the Baltic Sea, lead in the Greenland ice cap, and so on. The evidence is, however, selectively chosen and covers only those areas of pollution where little attention has been paid to control programmes. A different picture is painted if one turns to those successful areas of legislation involving often only small amounts of public expenditure.[...] Legislation, such as the Clean Air Act 1956 in Britain, has helped to effect pollution control. Despite a ten per cent increase in population and a seventeen per cent increase in energy consumption in the fifteen years following the Clean Air Act, there has been a steady reduction in smoke and sulphur dioxide emissions into the air over Britain. In many large towns this has provided the added bonus of increased winter sunshine. In Central London there has been a 50 per cent improvement in mean sunshine hours per day, the differential between previously less polluted areas such as Kew having narrowed considerably (Royal Commission on Environmental Pollution, 1971).

There have been, of course, more recent problems arising from pesticides, lead smelters, mercury poisoning and nuclear wastes. However, successful pollution control in the past makes it plausible that such problems are just as amenable to control as were the older forms of abuse. It might be argued that if chemical pesticides do become a real threat to economic growth through the pollution they cause, then alternative biological pest control (or even changes in the use of agricultural land) could offer substitute means of pest control without the same costs. Even without such dramatic changes, W. Beckerman (1974) claims that satisfactory pollution control programmes are well within the grasp of advanced industrial economies.

Simple Malthusian arguments have been countered in the main by historical reference to technological improvements and substitution. However, faith in technology as a saviour may well be unwise, for past experience (or at least this kind of interpretation of the past) is not necessarily a good guide to the future. The fact that the efficiency for generating electricity in the UK was about eight per cent in 1900 and is twenty-five per cent today is no guarantee of ever-increasing efficiency: indeed, the best possible practical efficiency is predicted to be around forty per cent. Furthermore, improvements in mining efficiency cannot be guaranteed to offset decreasing average grades of ores.[...]

Energy analysis of food production demonstrates even more dramatically diminishing returns from energy inputs. The example of maize (the most important grain crop grown in the United States, and ranking third in world production of food crops) illustrates the point: despite an increase in maize yields on United States farms from 34 bushels per acre in 1945 to 81 bushels per acre in 1970, the mean energy inputs increased from 0.9 million kcal to 2.9 million kcal. The total maize yield can be translated into energy equivalents so that in 1945 the maize yield was equivalent to 3.4 million kcal, and in 1970 to 8.2 million kcal. Hence, the yield in maize calories decreased from 3.7 kcal per fuel kilocalorie input in 1945 to a yield of about 2.8 kcal in 1970 (Pimentel *et al.*, 1973).[...]

The Malthusian argument has greatest relevance in relation to those resources that have least potential for expansion, especially those resources that can be loosely defined as providing rural amenity. There has, for example, been a huge loss of wetlands in America. The loss of rural land in England and Wales to urban development will, according to estimates by R. H. Best (1976), have increased threefold from 1900 to the year 2000. At approximately one per cent growth in urban land per decade, some 14 per cent of the total agricultural land will have been lost by the end of the century. Furthermore, conflicts in the UK National Parks over mineral extractions, water resources development, and the impact of an increasing number of holiday-makers, all illustrate the problems of maintaining amenity for an increased population with higher living standards. In this respect, the arguments for a stationary state put forward by John Stuart Mill make greater sense than many more recent arguments. He argued that the loss of diversity in nature and of wilderness resulting from the need to produce more and more food was undesirable.

A somewhat similar but less acceptable argument has recently been raised, with a good deal of favourable response, by F. Hirsch in *Social Limits to Growth* (1977). Hirsch makes a distinction between material goods such as food, which can be enjoyed irrespective of what other people are eating, and 'positional goods'. Positional goods can be enjoyed most if other people do not have access to them. The beautiful view from the front window is only enjoyed so long as there are not other houses in front enjoying the same view. The peaceful drive in the country only remains peaceful so long as there are not too many others searching for a peaceful drive. Now, according to Hirsch, social limits to growth exist because with increased growth competition moves increasingly from the material sector to the positional sector. As a consequence, the number of positional goods is fast decreasing in quantity and value.[...]

The similarity with the Malthusian position is not hard to see. It is advanced in such a way as to support policies that might maintain positional goods for the elite. It is also profoundly ideological in the sense of being an ascientific delusion. Like Malthus, little or no allowance is made for the creation of new 'positional' goods. The example of a peaceful ride in the motorcar is itself only possible given the invention of the car. New inventions are forever creating new positional goods whether they be motorboats, yachts, aeroplanes, hang-gliders, or whatever the latest plaything for the rich and leisured classes may be. Moreover, the potential for producing better ways of enjoying amenities for the great majority of people would not seem to be constrained by shortages of space or resources. There would be many ways of enriching the quality of life, access to privacy and sociability if greater effort were allowed to be put in this direction.

Other social limits to growth have also been put forward, such as the trend towards greater crime and violence, the psychological remoteness of industrial society, the institutional chaos and complexity of wealth creating activity involving multinationals, multiple unions, finance houses and government (Robertson, 1977/78). These problems and potential concerns are real enough, they certainly appear to be symptoms of a disease, but is the problem really related to economic growth itself or advanced capitalism and the monopoly power of big business? Surely none of these problems above need be a consequence of higher standards of living. Indeed there seems to be little hard empirical evidence that would support such an argument.

The economic technological fix position

The argument from this position is essentially that the scarcity of resources is governed by market price factors which influence the search for new resources, substitution, recycling and conservation measures. For instance, Beckerman argues that 'the market mechanism has hitherto usually ensured that, sooner or later, either increasing demand for materials has always been matched by increasing supplies, or some other adjustment mechanism has operated' (1974, pp. 34–5).

As already noted, optimists such as Kahn point out that known reserves are more a reflection of mining companies' search policies than of total resources. North Sea oil is another case in point.[...]During a twelve-month period in 1975–6, the known North Sea oil reserves in the British sector increased from 1,000 million tonnes to 1,350 million tonnes. There were twenty-four oil discoveries, nearly as many as in the previous five years; but towards the end of this period the rate of discoveries was falling off, due to cost inflation and strained finance (*Guardian*, 30 April 1976). Production costs, oil prices, finance and available technology are therefore of crucial importance in assessing known and estimated reserves.[...]

The energy economist, M. Posner (1974), has argued that North Sea oil, as well as deep-mined coal or oil shales, may be only marginally profitable if the price of oil falls to $5 per barrel at 1974 prices.[...]Moreover, while the availability of both oil and other resources bears some relation to demand, the actual market price of natural resources appears to have little to do with their physical scarcity, but much to do with monopoly positions.

The liberal economic view, as S. R. Eyre (1978) points out, plays down the importance of resource availability as a factor in the potential wealth and growth of a nation. With the growth of nationalism and separatism which has been taking place since the early twentieth century, there is every indication that resource endowment will be more likely to play a crucial part in the wealth of nations.

Unfortunately, the distribution of resources in general is very uneven. Non-ferrous minerals important for industry tend to be concentrated in Southern Africa, South-East Asia and China, the USSR and the western part of the Americas (Roberts, 1978). For example, in the case of phosphorus some 80 per cent of the world's output is used in the manufacture of fertilizers and there is no obvious possibility of substitution. Moreover, 75 per cent of the production of phosphate rock is confined to three countries – the United States, the Soviet Union and Morocco. Given the large domestic consumption in the two super-powers, Morocco, with 34 per cent of the world trade, has much influence over supply (and ultimately over the price) of phosphate rock. This situation suggests that the availability of a crucial resource will be determined not just by economic and technical factors but by the politics of a few countries in a monopoly position.[...]

The Marxist political economy position

This perpective, like the previous one, claims that resources and pollution control cannot be viewed apart from economic processes. However, unlike other perspectives, the question of resource scarcity and pollution control are linked to the organization of capital, the modes of production, and the power bases within society. Marx's answer to Malthus was a rejection of the idea of an absolute law dictating that there will always be more people on land than can be maintained from the available means of subsistence unless checked by famine, war, pestilence or artificial controls. Marx (1970, pp. 591–2) argued that every stage of economic development has its own law of population. It was the surplus population of unemployed workers that, in the capitalist system, led to poverty and the appearance of over-population. The creation of the surplus population was not a product of resource scarcity but of the capitalist mode of production. Capitalist accumulation decreased the proportion of variable capital to constant capital. The amount of labour is influenced by the amount of variable capital. Consequently, if the variable capital fails to rise with population growth due to capital accumulation, then a surplus population is created. In the words of Marx:

The fact that the means of production, and the productiveness of labour, increases more rapidly than the productive population, expresses itself, therefore, capitalistically in the inverse form that the labouring population always increases more rapidly than conditions under which capital can employ this increase for its own self-expansion. (1970, p. 604)

Even for those who are employed, low wages, influenced by a surplus population of unemployed, create inadequate demand (rather than need) for food and other products. There are no physical constraints on production, but the economic organization of society is responsible for the lack of demand among the poor. Engels rejected the 'limits' argument of Malthus as follows:

Too little is produced, that is the cause of the whole thing. But *why* is too little produced? Not because the limits of production – even today and with present-day means – are exhausted. No, but because the limits of production are determined not by the number of hungry bellies but by the number of *purses* able to buy and to pay. Bourgeois society does not and cannot wish to produce any more. The moneyless bellies, the labour which cannot be utilized *for profit* and therefore cannot buy, is left to the death rate. (Meek, 1971, p. 87)

The problem of effective demand for resources is of crucial importance in agriculture. There is, for instance, at present no *absolute* shortage of grains in the world, and yet there are acute *regional* shortages. According to Jean Mayer of Harvard University, the same amount of food consumed by 210 million Americans could adequately feed a population of 1.5 billion people at the Chinese level of diet (Power and Holenstein, 1976). N. Eberstadt (1976) has pointed out that the food production *per capita* rose by nine per cent in fifteen years after 1960, that there is more than enough food adequately to feed the world population, and yet millions of people still starve (see also Rothschild, 1976). In Africa barley, beans, cattle, peanuts and vegetables are exported despite the fact that malnutrition is worse in Africa than on any other continent (Lappe and Collins, 1977). Poverty is due to maldistribution of resources (both internationally and within nations), and not to the physical limits of producing the resources themselves. Indeed, given the situation of maldistribution of wealth, land and economic opportunity, the introduction of more productive agriculture (as in the case of the Green Revolution) can lead to a *worse* distribution of wealth and a lower effective demand for agricultural resources. [...]

Scarcity, far from being a natural phenomenon or a state resulting from economic growth, is managed in such a way as to maintain a demand that exceeds supply and consequently results in a handsome profit. Only in such terms can one understand the waste resulting from the dumping of surplus milk by the Americans in the 1960s – or the cut-back in wheat acreages from 120 million to 81 million acres in the US, Australia, Argentina and Canada during 1968–70, the stock-piling of beef and butter mountains in the EEC which are sold off cheaply to the USSR, or the throwing away of vast quantities of French apples during the autumn of 1975. In the summer of 1977 the US Administration were again planning to reduce the growing surplus of world wheat by allowing millions of acres of productive farm land to become fallow. Despite the persistence of world hunger, the surplus of US grain was becoming unmanageable; its reduction would help to halt the fall in wheat prices, and so protect the American farmer. [...]

The question of whether or not resource scarcity and waste will be a problem in the future is therefore inextricably linked to further questions concerning what type of social relations will exist in the future. Contradictions leading to pollution and resources depletion are only one type of problem that capitalism faces. [...] Advances in the micro-electronics industry, which in Britain alone 'threatens' to release several millions from the 'labour market' in the late twentieth century are consequently greeted with widespread alarm and despondency. Over-production and time-saving techniques can bring about undesirable social problems in a market economy; they may lead to depression and economic crisis.

The Marxist position, like the economic/technological fix position, has tended to ignore the importance of resource endowment in the development of wealth. While social relations of production and economic organization in general are obviously important there is, as Eyre (1978) has pointed out, a great geographical variation in resource endowments which certainly constrain the path towards a higher standard of living. There has always been a clash between Marxist and Malthusian theorists about the relationship between population growth and poverty. Nevertheless, for pragmatic reasons a population policy in countries of low resource endowment in relation to population size makes sense. To some extent China's policy on population has from time to time (especially between 1954 and 1958, 1962 and 1966 and since 1969) recognized this argument, even though China herself is reasonably well endowed with resources in comparison to other Asian countries such as India. Small family size and later marriage in China have, on the other hand, usually been encouraged for reasons of family health and prosperity.

Recently some Marxists have argued that scarcity of mineral reserves at a cheap price threaten advanced capitalist states. A. Gedicks (1977) claims that Marx recognized the importance of low cost resources in order for the process of capital accumulation to continue. [...] Gedicks goes on to argue that scarcity of indigenous reserves will compel the United States to seek out stable supplies of cheap resources against the trend of growing nationalist tendencies.

The main example that Gedicks uses to support his thesis is that of American imperialism in Chile, the object of which was to protect supplies of cheap copper. Prior to the Allende Government in Chile, copper production was geared to the requirements of the American market. When these mines were nationalized, the US government began an economic and political campaign to undermine Allende's government. Eventually, with United States help, the Allende Government was overthrown and a military dictatorship established.

The threat of nationalism and loss of control over physical resources, albeit important, may not be quite as serious for the growth of capitalism as Gedicks implies. There may well be a shift of manufacturing to the less developed countries while the advanced countries shift increasingly towards becoming 'service' economies. K. Kumar (1979) argues that this trend is already taking place in Britain. Gone are the days when Britain was the major manufacturing exporter. In 1870 her exports claimed 40 per cent of the world trade in manufactures. In 1976 this figure had fallen to below 9 per cent. On the other hand, Britain is second only to the United States in the world trade of 'invisibles' (services).

Nevertheless, aside from questions as to whether or not the capitalist social order is about to break down, the Marxist account offers a plausible explanation of why there is still resource scarcity for large numbers of the population. Starvation and shortages of basic material requirements owe more to the influences of capitalist ownership and consequent imperialism and suppression of the working class, than to any real physical constraints on providing for those requirements. To this extent the pessimistic conclusions of the Malthusian standpoint are ill-founded and based upon ideological rather than scientific understanding. The economic/technical fix position is also ideological in the sense of being limited to supply and demand equations without considering the influence of capitalists in manipulating both in the interests of capital accumulation rather than of meeting human needs.

Predictions for the future

In a review of future studies, S. Cole, J. Gershuny and I. Miles (1978) discovered no less than sixteen major reports on the predicament of mankind which were published between 1965 and 1977.[...]A more serious appraisal of these futuristic studies is appropriate here. This can be accomplished by evaluating the same three views that were considered in the limits to growth debate.

The neo-Malthusian position

Some of the predictions have started from Malthusian assumptions about the limits to growth. R. L. Heilbroner's *An Inquiry into the Human Prospect* is a good example. He identifies three major problem areas which are externally generated and threaten "the human prospect". They are population, nuclear war, and the physical limits of the environment to the sustainment of growth. All of these problems he lays at the door of science and technology. The population problem is a consequence of "a science-induced fall in death rates" (1975, p. 56). The possibilities of nuclear war and environmental catastrophe are also linked to developments in science and technology. He argues that industrial developments induce similar social consequences, whether it be under capitalism or socialism.

Having decided that industrial growth will have to slow down because of the three limiting problems outlined above, Heilbroner then goes on to discuss the capacity of different political institutions to adjust. Acceptance of the concept of limits to growth produces a prediction for the future which he believes will probably require strong centralized political power. He doubts whether 'human nature' would allow for peaceful and organized changes in life-style.

E. J. Mishan (1977) falls into both the Malthusian and technological determinist trap. He argues that the predictions of Malthus and Ricardo were premature rather than wholly wrong. Recent economic and technological growth has threatened humanity because of consequences arising from improved transport and communications; rapid innovations; weapons technology; urban size and concentration. Once again Mishan attributes these dangers to economic and technological factors rather than to any underlying characteristics of the social order. Like Heilbroner, Mishan believes that the momentum of economic growth and technological development is only likely to be controlled by a totalitarian state.

C. Taylor (1978) also accepts that population, resource scarcity and pollution threaten to limit growth, and as a consequence it will be necessary to move towards a Byzantium-type steady state. He too believes that this transition will probably come about under authoritarian regimes. The problem a capitalist society faces, he argues, is that inequalities are only bearable when there is economic growth. Stagnation would induce tension between privileged and less privileged groups. This could only be contained by more authoritarian regimes or even dictatorships which would either uphold the inequalities or do something about them.[...]

These three predictions from Heilbroner, Mishan and Taylor bear close similarities. They can all be dismissed on the grounds of holding false premises derived from the Malthusian perspective and technological determinism.

Some of the radical views of the future which search for liberation from industrialization also suffer from misinterpretation of the nature of the crisis which would bring about this future. In the writings of I. Illich (1973) and T. Roszak (1972) in America and E. F. Schumacher (1973), J. Robertson (1977/78) and a good many of the radical technology theorists in Britain, there is a common assumption that a crisis results from the independent development of science and technology. As K. Kumar comments, "in this account the crisis is the inevitable product of the long-term tendency of industrialism towards the large-scale, greater centralization, a finer specialization and division of labour, and the replacement of human labour and human skill by a resource-consuming machine technology" (1979, p. 15).

The economic technological fix position

The next series of predictions are those based upon premises derived from the economic/technological fix perspective. Futurologists such as Kahn (1976), D. Bell (1973) and P. F. Drucker (1969) hold an essentially business-as-usual position. They are optimistic about the industrial future. The way the future is likely to differ from the present is considered in terms of the types of technologies that will be developed in order to sustain growth. The shift will be from conventional manufacturing to high technology such as nuclear fission and fusion. Widespread automation, further developments in transportation, aerospace, telecommunications, computerization and even the possibilities of space colonization will bring us into the hyper-industrial estate. The problem facing governments is how to make the transition as smooth as possible (see Robertson 1977/78). [...]

In many respects the hyper-industrial future based upon liberal economic principles is a likely prospect because it is a projection of numerous tendencies that exist in the advanced capitalist nations today. The question that needs to be raised when assessing its plausibility is whether or not the contradictions of capitalism which today give rise to problems of pollution, resource scarcity and risks from Flixborough, Seveso and giant oil-spillage disasters can be overcome without both a change in social order and a change in the path of technological development. Given the record of capitalist countries in tackling these problems, and the tendency towards greater risks in industrial development, the prospects of this prediction occurring are grave indeed. Together with other contradictions of capitalism, such as poverty in the midst of plenty and high levels of unemployment, it is not unlikely that movements toward an alternative path of development will achieve some success.

H. Stretton (1976) has described three possible directions for the future based upon reformist policies. The first two directions involve a shift to the right. The first is that a reactionary class-ruled regime which adopts preservationist policies along strictly Malthusian lines will take control. Inequalities will increase and the state will become more authoritarian: the death penalty, strict population control policies and penal colonies may be some of the instruments of social control.

His second prediction suggests that political influences slightly to the right will bring about environmental reforms which mildly increase divisions and inequalities within society. Just as the better off have benefited most from environmental planning and pollution control in the past, so they will continue to do so in the future. [...]

Stretton's third prediction involves a political shift to the left; left-wing parties gain in strength from the "cost-of-living troubles and rising inequalities ...urban and environmental troubles, quality of life issues, local and community initiatives, business scandals and property prices and the problems of inflation" (1976, pp. 97–8). The traditional Left and environmentalists, who were estranged from each other in the late 1960s and early 1970s, would find common ground in campaigning against environmental dangers to basic necessities. While private enterprise continues, its activities are increasingly governed by labour interests. Instead of the capitalist class controlling the state, the reverse process would gradually take place. [...]

Stretton is right to point out alternative paths of reform under capitalism. However, his predictions flirt with the possibilities of Malthusian constraints without assessing their likelihood. It is also doubtful whether the type of

Leftist reforms which he envisages could take place while maintaining the capitalist economy intact. The history of labour reformist intervention to date does not suggest that political reforms alone will be sufficient to erode inequalities or the power of the capitalist ruling class (see Westergaard and Resler, 1976). There is obviously a spectrum of possibilities between a *laissez-faire* economy and a socialist state where private ownership plays an insignificant role. However, the most important areas for change are not the state and policy-making institutions, but control over the forces of production. It is only when product innovation, pollution controls and resources utilization by firms are taken out of the control of management and entrepreneurs whose prime aim is to maximize profits that true progress can be made towards a healthier environment for all.

The Marxist political economy position

The last type of prediction stems from a Marxist viewpoint. It assumes that planning and technological choice and economic development will come under the control of the workers and the local community rather than the financial magnates of industry. Under both state and local control, economic development could be planned to improve environmental working and living conditions, and wastage would be reduced by avoiding unnecessary competition. Food, shelter, energy and mineral resources would be regarded as precious and not to be squandered as occurs today under the distortions of a market economy.[...]Individuals would become freer to travel with improved public transport, cycling and pedestrian conditions, and the wastage from half-empty cars would be avoided.

Today polluting industries, traffic hazards and other environmental disturbances are unevenly distributed so that the better off have prime access to higher quality environments. The rich and dominant class have been able to manipulate political institutions so that planning reinforces their narrow needs. In the social future political institutions and planning would serve a wider spectrum of interests. Industrial democracy and choice of products according to need rather than profit would ensure a more efficient use of resources. The struggle by the Lucas Aerospace shop stewards for their corporate plan would be remembered as a milestone in the creation of a new future where there was a real freedom of choice for the majority of people. People and politics would be in command rather than the demands of capital accumulation for its own sake.

Whether or not the socialist prediction would be constrained by the availability and price of physical resources would still be debatable, especially in countries with few resources. Exploration and development of physical resources would not, however, be determined by profit implications, and there would be less unnecessary exploitation for products which meet 'false needs'. The likelihood of this third prediction occurring is in doubt, but one can claim that its development would ensure a safer and more fulfilling future. Whether or not such changes are realized before the development of further major wars or before some major disaster from bio-chemical pollution or radiation remains to be seen.

Conclusions

In a typical liberal and pluralistic account, environmental problems arise at various stages during the process of industrialization. Consequent strains lead to the development of social movements. The state, pressure groups, and the public react to these problems and, depending upon the nature of the political system, there are varying degrees of consensus and conflict before acceptable solutions are found. In the democratic capitalist economies dominant social science and policy analysis implicitly deny that conflicts of material interest play a significant part in generating and resolving social and environmental problems. 'Scientific rationality' is, however, closely integrated with a system of capitalist production (see Gorz, 1976(b)). Hence cost–benefit analysis and technology assessment are legitimate and respected forms of policy analysis whereas the activities of workers' organizations aimed at transforming work organization and industrial production are regarded as political rather than scientific.

If the policy outcome of cost–benefit analysis, behavioural studies, environmental impact assessment and technology assessment is shown to be politically biased – favouring middle-class owner-occupiers, for example – then this is often claimed to be due not to a political bias of the ideas governing policy analysis themselves but an abuse of these ideas. The development of ideas and theories governing policy and the understanding of social movements, pressure groups, the behaviour of firms and the growth of the state are assumed to be free from ideology.[...]

A use/abuse model of scientific knowledge is supported in the main by historians and philosophers of science. The development of science is seen as following an independent and self-governing path. No significance is granted to the influence of external economic and social conditions on the nature and development of science.

Although there is a semblance of explanatory logic in pluralist, functionalist, behavioural and neo-classical theories, which dominate the social scientific analysis of environmental problems and policy, they can all be shown to be defective, in as much as they neglect material interests. Systems of ideas, whether they be idealist or economic, which are divorced from the real material world serve to obscure the true basis of environmental and social problems. Those who support systems of thinking that are independent of the social organization of production and claim them to be ideologically neutral are profoundly misleading.[...]

In the social sciences the production of knowledge is distorted by specialization and fragmentation of academic disciplines, and studies which play a functional role of supporting the *status quo* are particularly promoted within academic and policy institutions. The result is that the 'ruling ideas' are, as Marx asserted, the ideas of the 'ruling class'. The solution to environmental problems depends therefore not merely upon reforming the science and technologies available but transforming the social relations of production – production of policy, technology, and knowledge itself. Resistance to nuclear fast-breeder reactors, to dangerous industrial developments, to impersonal environmental planning, is not merely a matter of improving expertise and the scientific basis of decision-making. It is a matter of claiming the right for all to control what is produced and what is planned.

References

Barnett, H. J. and Morse, C. 1965: *Scarcity and growth: the economics of natural resource availability.* Baltimore: Johns Hopkins University Press.

Beckerman, W. 1974: *In defence of economic growth.* London: Cape.

Bell, D. 1973: *The coming of post-industrial society: a venture in social forecasting.* New York: Basic Books.

Best, R. H. 1976: 'The changing land use structure of Britain.' *Town and Country Planning* 44, 171–6.

Chapman, P. 1975: *Fuel's paradise: energy options for Britain.* Harmondsworth: Penguin.

Cole, S., Gershuny, J. and Miles, I. 1978: 'Scenarios of world development.' *Futures* 10, 3–20.

Connelly, P. and Perlman, R. 1975: *The politics of scarcity in resource conflicts in international relations.* London: Oxford University Press.

Drucker, P. F. 1969: *The age of discontinuity.* New York: Harper and Row.

Du Boff, R. B. 1974: 'Economic ideology and the environment.' In Van Raay, H. G. T. and Lugo, A. E., editors, *Man and environment Ltd.* The Hague: Rotterdam University Press, 201–20.

Eberstadt, N. 19 February 1976: 'Myths of the food crisis.' *The New York Review of Books* 23, 32–7.

Ehrlich, P. R. and Ehrlich, A. H. 1970: *Population, resources, environment: issues in human ecology.* San Francisco: W. H. Freeman and Co.

Engels, F. 1964: *Outlines of a critique of political economy* (first published 1844). New York: International Publishers.

Eyre, S. R. 1978: *The real wealth of nations.* London: Edward Arnold.

Foley, G. 1976: *The energy question.* Harmondsworth: Penguin.

Gedicks, A. 1977: 'Raw materials: the Achilles heel of American imperialism?' *The insurgent sociologist* 7, 3–13.

Gorz, A. 1976: 'On the class character of science and scientists.' In Rose, H. and Rose, S., editors, *The political economy of science. Ideology of/in the natural sciences.* London: Macmillan, 59–71.

Heilbroner, R. L. 1975: *An inquiry into the human prospect.* London: Calder and Boyars.

Hirsch, F. 1977: *Social limits to growth.* London: Routledge and Kegan Paul.

Illich, I. D. 1973: *Tools for conviviality.* London: Calder and Boyars.

Jevons, W. S. 1965: *The coal question: an inquiry concerning the progress of the nation and the probable exhaustion of our coal mines* (first published 1865). Third edition, editor, Flux, A. W. New York: Kelley.

Kahn, H., Brown, W. and Martel, L. 1976: *The next 200 years – a scenario for America and the world.* New York: Morrow.

Kumar, K. 1979: 'Thoughts on the present discontents in Britain: a review and a proposal.' To be published in *Theory and Society.*

Lappe, F. M. and Collins, J. 1977: *Food first: beyond the myth of scarcity.* New York: Houghton Mifflin Co.

Leach, G. 1976: *Energy and food production.* Guildford, Surrey: IPC Science and Technology Press.

Malthus, T. R. 1970: *An essay on the principle of population* (first published 1798). Harmondworth: Penguin.

Marx, K. 1970: *Capital*, Volume 1 (first published 1867). Moscow: Progress Publishers.

Meadows, D. H., Meadows, D. L., Randers, J. and Behrens III, W. W. 1972: *The Limits to Growth.* New York: University Books.

Meek, R. L. 1971: *Marx and Engels on the population bomb.* Berkeley, California: Ramparts Press.

Mishan, E. J. 1977: *The economic growth debate: an assessment.* London: George Allen and Unwin.

Page, W. 1973: 'The non-renewable resource sub-system.' In Cole, H. S. D., Freeman, C., Jahoda, M. and Pavitt, K. L. R., editors, *Thinking about the future: a critique of the limits to growth.* London: Chatto and Windus, 33–42.

Pehrson, E. W. 1945: 'The mineral position in the United States and the outlook for the future.' *Mining and Metallurgy Journal* 26, 204–14.

Pimentel, D., Hurd, L. E., Bellotti, A. C., Forster, M. J., Oka, I. N., Sholes, O. D. and Whitman, R. J. 2 November 1973: 'Food production and the energy crisis.' *Science* 182, 443–9.

Posner, M. 1974: 'Energy at the centre of the stage.' *The Three Banks Review* 104, 3–27.

Power, J. and Holenstein, A. 1976: *World of hunger – a strategy for survival.* London: Temple Smith.

Roberts, F. 6 April 1978: 'And now – a resources shortage?' *New Scientist* 78, 16–17.

Robertson, J. 1977/78: 'Towards post-industrial liberation and reconstruction.' *New Universities Quarterly* 32, 6–23.

Roszak, T. 1972: *Where the wasteland ends.* London: Faber and Faber.

Rothschild, E. 1976: 'Food politics.' *Foreign Affairs* 54, 285–307.

Royal Commission on Environmental Pollution 1971: *First report.* Cmnd 4585. London: HMSO.

Schumacher, E. F. 1973: *Small is beautiful: a study of economics as if people mattered*. London: Blond and Briggs.

Shenfield, A. 11 August 1977: 'Energy and doomsayers.' *The Daily Telegraph*, 16.

Stretton, H. 1976: *Capitalism, socialism and the environment*. Cambridge: Cambridge University Press.

Taylor, C. 1978: 'The politics of the steady state.' *New Universities Quarterly* 32, 157–84.

Westergaard, J. and Resler, H. 1976: *Class in a capitalist society*. Harmondsworth: Penguin.

Index